Functions: selection and mechanisms

SYNTHESE LIBRARY

STUDIES IN EPISTEMOLOGY, LOGIC, METHODOLOGY, AND PHILOSOPHY OF SCIENCE

VOLUME 363

For further volumes:
http://www.springer.com/series/6607

Philippe Huneman
Editor

Functions: selection and mechanisms

Springer

Editor
Philippe Huneman
IHPST (CNRS/Université Paris I Sorbonne)
Paris, France

ISBN 978-94-007-5303-7 ISBN 978-94-007-5304-4 (eBook)
DOI 10.1007/978-94-007-5304-4
Springer Dordrecht Heidelberg New York London

Library of Congress Control Number: 2012956316

Printed on acid-free paper

Springer is part of Springer Science+Business Media (www.springer.com)

Acknowledgements

This book has been possible because of the activity of a research group at the Institut d'Histoire et de Philosophie des Sciences et des Techniques (CNRS/Université Paris I Sorbonne), which addressed philosophical issues about functions in the framework of two funded projects (CNRS and ANR) entitled "Functions and functional explanations in biomedical sciences." Some of the contributors authors actively contributed to seminars or workshops held by the group.

I warmly thank all of them here.

The research group greatly also benefited from the input of other philosophers closely associated, among them Tim Lewens, Elodie Giroux, Françoise Parot, Denis Forest, Peter McLaughlin, Francesca Merlin, Christophe Malaterre and Thomas Pradeu.

Along those years, Marie-Claude Lorne has been a constant stimulation and inspiration for our research. She tragically left us in 2008. This book is dedicated to her.

Contents

Part I
Biological Functions and Functional Explanations: Genes, Cells, Organisms and Ecosystems – Functions, Organization and Development in Life Sciences

Introduction

Philippe Huneman

Abstract This introduction presents general issues about functions and functional explanations, and the frameworks within which they had been handled by philosophers since four decades. It sketches the current state of the art, indicates areas in biology, cognitive science, philosophy of science and metaphysics, that call for further investigations about these topics, therefore explaining the project of this volume. It ends by describing the general articulation of the book and providing an overview of the contributions that the reader will find here.

1 The Theories of Function and the Current Issues

This collection of chapters aims at reflecting upon the metaphysics of function and the various problems that functional explanations raise. The question of function and functional explanations has certainly been extensively dealt with by philosophers of biology, as well as by philosophers of action and philosophers of mind. Since the early 1970s, the concept of function, as used in biology, psychology, and related disciplines, has indeed continuously been under philosophical scrutiny. The origin of these discussions is to be found in the two papers published by Larry Wright and Robert Cummins in 1973 and 1975 respectively. These papers renewed the debate, with two innovative analyses going in different directions. The *etiological theory of functions* (or "selected effects" functions (Neander 1991), or "teleofunctions," or "proper functions" (Millikan 1984)), which stems from Wright's paper, holds a realist concept of function and, in the case of Wright himself, aims at a unified theory of artifacts and biological entities. Against this realistic claim,

P. Huneman (✉)
IHPST (CNRS/Université Paris I Sorbonne), Paris, France
e-mail: philippe.huneman@gmail.com

P. Huneman (ed.), *Functions: selection and mechanisms*, Synthese Library, Studies in Epistemology, Logic, Methodology, and Philosophy of Science 363,
DOI 10.1007/978-94-007-5304-4_1,

Cummins defended a concept of functions (as "causal role" in a system) that makes them relative to an explanatory strategy, which has to define a system within which the functional item is embedded. Both acknowledge that "function" is a concept used in some explanations, but they diverge from the first step because the etiological account thinks that the function of X being Y explains the presence of X, whereas for the causal-role theorist, the function of X being Y explains or contributes to an explanation of the general proper activity of a system which includes X.

The etiological theory faced several objections and was refined through numerous debates in the two last decades (e.g., Godfrey-Smith 1993; Kitcher 1993; Buller 1998, etc.). Similarly, the causal-role theory of functions increased in sophistication, as researchers were finding new patterns of explanation that made use of it for particular cases, for example, when Amundson and Lauder (1994) emphasized its major role in functional morphology. Given that the two analyses seemed to be adequate in distinct areas of biology, and that, moreover, those two analyses accounted for different functional ascriptions of a same item that could be met in one given field of this science, important papers such as Kitcher (1993), Millikan (2001), or Godfrey-Smith (1993) considered ways of articulating the two approaches and many authors subscribed to some sort of pluralism. So even though the two concepts rest on opposite assumptions – especially, as mentioned above, about what the explanandum of a functional explanation should be – they situate sets of nuanced views rather than two monolithic positions; those two sets are such that in each of them, pluralist positions are easy to be found.

In broad outline, the two following claims about functions make up the general framework for the discussion: (a) Functions are generally implemented in mechanisms; (b) functional explanations in biology have an essential relation with natural selection. Each main account of functions emphasizes one aspect and downplays the other. For instance, when one says that (F) "the function of the vertebrate eye is seeing", this relates to two sets of facts at the same time: there is a vision mechanism, quite sophisticated, involving at least the eye, nerves, and the brain – and the eye is the result of a complicated process of natural selection across vertebrates, first sketched by Darwin in the Chapter 5 of the *Origin of species*, and lastly modeled by Nilsson and Pelger (1994). Understanding what functions and functional explanations are therefore requires one to take a stance regarding these two aspects. An etiological theorist will claim that the main aspect is the natural selection, which accounts for the explanatory role of the statement (F) regarding the presence of the eye, and it can account in the same manner for functions of items that involve a very crude mechanism, for example, the fur color of tigers. Yet there is also a thin mechanism here (the fur makes the tiger match its surroundings, so prey has fewer chances to see it), even if it is far less complicated than the vision mechanism. Inversely, the causal-role theorist will emphasize the first aspect, the mechanism, and on this basis will account for the explanatory role of (F) regarding the general perceptual ability of the vertebrate. Then, as soon as items involve some mechanism, the embedding of such a mechanism in a large system will ground a functional statement understood in causal-role terms, even if selection has been controversially acting, or if the item has been demonstrated to be the outcome of drift or a mere byproduct of selection.

Pluralist positions – namely, etiological theory that, within the continuum of etiological positions, do not tie the whole account to the fact of selection (e.g., Buller 1998; Kitcher 1993) or causal-role theorists who admit that mechanisms are there because of natural selection, to which they owe many features – will be more likely to make room for both aspects, and tackle the issue of their articulation (in general, and in each distinctive field). Given that reflections on functions and functional explanations, in philosophy and in life sciences at least, became so sophisticated and allowed for many pluralist stances to be founded on both sides, this urges a renewed understanding of the possible relations between (a) and (b).

The title of the present collective book is therefore: *Function: selection and mechanisms*. This book intends to cast new light upon the two rough claims (a) and (b) by confronting them with scientific developments in biology, psychology, and recent developments in metaphysics.

In effect, various developments in biology, engineering, cognitive sciences and philosophy of science compel us to think that, notwithstanding the degree of sophistication reached by philosophical theories of function at the end of the 1990s, issues around functions have not yet been solved. The framework of the philosophical understanding of function and functional explanation has been actually changed by the following advances.

- A. Regarding philosophy of science, strictly speaking, a new position has been defined in the context of the causation and explanation debates, which has been called "the mechanistic view of science" (Machamer et al. 2000). This view holds that science does not formulate laws or pick out causes but mainly describes mechanisms – and its proponents make explicit in their papers what this rough characterization involves. Mechanisms are supposed to only involve specific "entities" with specific "activities", all of which are sufficient to account for the way mechanism functions, and yields the phenomena to be explained as its outcome. Interestingly, the main application of this concept has been within the biological sciences, for example, molecular biology (Darden 2006) and neurosciences (Craver 2008). Philosophers debate about whether it correctly captures the metaphysics of science or only captures the activities of scientists. Yet, since this approach views the objects of science as mechanisms made of entities with activities, and those mechanisms function, one could suggest that those activities are the functions of the entities. Hence, the mechanistic view of science raises conceptual issues about the meaning of function and the role of functional explanations in the depiction and uncovering of those mechanisms.
- B. In the cognitive sciences, new developments such as "situated cognition" or "embedded cognition" (e.g., Shapiro 2010), challenging the classical cognitive and connectionist viewpoints, offer new insights on what it means to have a function. Such is also the case with the refined analyses of explanations in neurology and in cognitive sciences that have been produced recently by various philosophers (e.g., Bechtel and Richardson 1993; Bechtel and Abrahamsen 2005; Craver 2001), often in connection with the "mechanistic" philosophy of science just mentioned. In many cases, the notion of function relies upon some understanding

of the mechanisms at stake in the cognitive devices and which scientists intend to grasp. This could be taken in favor of the simplest causal-role theory of functions; however, given that the very meaning of "mechanism" has been reworked, the consequences of these analyses are not so clear-cut. A new understanding of what mechanism should mean in the field would therefore impinge on what functions are and what they explain.

- C. In biology:

  - All etiological accounts of function share a general appeal to natural selection in order to make sense of the explanatory force of functional statements. However, what selection is and how it is ascribed is often not detailed in these accounts. For instance, Wright (1973) was very general and equated selection and choice as two modes of the same selection process, Millikan (1984) had a very idiosyncratic redefinition of selection, etc. Philosophers of biology sometimes saw the importance of being clear about what selection is for the function issue, as exemplified by the use of Sober's selection for/selection of difference in this context (Enç and Adams 1992). But in the last decade specifically, many issues have been raised about natural selection. Some may be too metaphysical to really impinge onto the function debate (e.g., whether selection is a cause or a statistical outcome (Walsh et al. 2002; Matthen and Ariew 2009; Lewens, 2009; Huneman 2012) or what selection actually causes (e.g., Sober 1995; Neander 1988). But some issues may be more directly relevant to biologists and therefore have consequences upon what functional ascriptions and explanations are.
  - C1. Lewontin (1970) had shown that any set of entities exhibiting *variation* over *heritable* properties causally related to differential reproduction (fitness) seemed to be potentially undergoing natural selection – and these seemingly necessary and sufficient conditions for natural selection were accepted for a long time by philosophers. Yet Godfrey-Smith (2009) has shown that natural selection is not easily captured in terms of necessary and sufficient conditions. In particular, there is a philosophical debate about whether heritability really is needed to define natural selection or not. *Evolution* by natural selection needs inheritance, according to Brandon (2010), but not natural selection itself. Additionally, the whole idea of fitness came unto scrutiny recently (Abrams 2009; Bouchard 2009, who challenged the whole frame of evolutionary population genetics.) Moreover, recently we saw the development of a multilevel selection paradigm for explanations of issues like cooperation (Damuth and Heisler 1988; Sober and Wilson 1998), evolutionary transitions toward individuality (Michod 1999; Okasha 2006; Griesemer 2000), or genomic conflict (Burt and Trivers 2006). Given that the etiological theory of function defines function by an appeal to selection, it will now be crucial for philosophers to confront multilevel selection in their conceptions of "selected effects" function, where implicit selection was usually assumed to operate either at the level of the gene or of the organism (Huneman 2013).

- C2. There is a growing discussion about the need to "extend" the classical evolutionary theory (e.g., Pigliucci 2007; Pigliucci and Müller 2011), stemming from the synthesis of Mendelian genetics and Darwinian transformism, essentially using population genetic models such as the ones designed by Fisher, Wright, and Haldane in the 1930s. From this viewpoint, the scope of selection may not be globally encompassing, especially, the variations upon which selection acts may have been more sophisticated than mere allelic mutations and recombination. In this view, the mechanisms producing variation, in a regular and systematic way, thereby have a crucial role in evolution, or at least macroevolution. Structuralists like Goodwin or Hall had, for a long time, argued that the important features in evolutionary long-run history and especially the commonalities of some forms and process across very distant phyla, may not be explainable by selection (e.g., Amundson 2005, for a short historical sketch). Current developmentalists (e.g., Raff 1996; Carroll 2005; Gilbert 2003, etc.) do not always undermine selection in such a hard way, but clearly, they advocate the role of other processes – acting within the stage of variation and often at the level of organisms rather than genes – in the shaping of living traits. If the role of selection in evolution regarding the explanation of diversity and even adaptation is to be reconceived, an account of functions based on natural selection, such as the etiological account, may become less accurate or at least less systematically valid for biology. It might mean a reinforcement of the causal-role theory; but it may also call for a more distinct reconception of what are functional traits functions of behaviors, etc. – a reconception which is likely to include development and organismal activity within the account.
- C3. Besides, ecology and evolution entered a new relationship. It has often been claimed that the evolutionary Modern Synthesis left ecology aside (e.g., Kingsland 1985), because it was centered on population genetics, which mainly targets one population of one species, whereas ecology considers sets of population of several species. Recently, we witness various attempts to synthesize ecology and evolution, be it in the context of niche-construction theory (Odling-Smee et al. 2003), in a reconception of the basics of ecology (Ginzburg and Colyvan 2004), or in the rise of metacommunity ecology (Leibold et al. 2004), especially in the form of the neutral theory of ecology (Hubbell 2001). Thereby, it makes it all the more important to understand functional explanations in ecological contexts, whereas the bulk of philosophical work has been centered on evolution.

– D. In metaphysics, functionalism has always been defined with a reference to multiple realizations (e.g., Putnam 1967; Fodor 1974). This constituted an important background for what philosophers meant by talking of functions. Functional properties were especially conceived of as a relation between a type of input and a type of output, the nature of whatever played the role of this relation being somehow irrelevant, and possibly infinite; this is the famous hyperbole by

Putnam, saying that even a chunk of Swiss cheese could think if it were exhibiting the appropriate functional correspondences.

Yet recently, coming from the philosophy of mind, attention has been paid to what "realization" means exactly (Shapiro 2000; Polger 2004, 2007; Gillett 2003), and philosophers emphasized difficult issues implicit in the very meaning of "realization" itself. This implies that, if philosophers still think of functions in terms of realization – for example, when they say that the same functional properties are realized by various possible processes – they will have to precisely make the metaphysical stance they adopt. Some of the stances, for example, do not entail Putnam's weird Swiss cheese consequence, because they restrict the ontological class of potential role-occupiers (e.g., Block 1997). That is the reason why the way one handles such issues about "realization" bears important consequences upon the very idea of function implied by functionalism, and finally on the concept of "function" in general. Granted, the "function" of functionalist philosophers of mind is not the "function" of behavioral ecologists, captured by philosophers who support the etiological theory of functions; however, as it is attested by the example of functions like seeing or storing or transmitting information, they are sometimes intended to capture the same core fact.[1] Therefore, the original issues about functionalism and functions in mind are relevant to a general questioning upon functions and the compared value of etiological and causal-role accounts.

Thus, in philosophy of science, cognitive and neuroscience, recent debates about evolutionary biology and ecology, and in the metaphysics of realization, are involved important consequences about the concept of function and functional explanation, that must in the end affect the traditional theories of function, even in their most sophisticated form. The chapters edited here intend to meet the challenge that this new scientific and philosophical context raises. They present and discuss issues on functions and functional explanations that have arisen recently, although not all the challenges listed here will be addressed.

2 Position and Structure of This Book

Such a collection has the double purpose of revisiting the sources of the debates and of presenting current investigations which show the complexity of the issues involving functional explanations in the various sciences. It includes papers both by authors of seminal papers in the controversies and by recent researchers who investigate the questions by adopting new perspectives. Thereby, it makes no a priori assumptions about the scope of functional explanations, and it touches upon several very different scientific domains. A possible overview of theories of function and practices of functional explanations is likely to be drawn from the whole book, but

[1] One can check out the table of function views by Polger (2004).

nothing has been done to hinder the tensions between rival approaches or just the divergence between consequences that one can draw from the exploration of different scientific areas.

An important philosophical question rising from a reading of the wide literature devoted to functions and functional explanations concerns the very nature of a philosophical account of functions. Like any philosophy of science project, understanding functions may be either a descriptive project – making sense of what scientists are doing with their functional ascriptions – or a normative one – determining the true nature of "function," and then dismissing these cases in the sciences that do not match it as non-genuine cases of functions. The latter project is more compelled to being somehow monist (function means one thing) than the former, which may by nature accommodate some pluralism since science has various legitimate modalities. Clearly, there is a continuum between those positions, especially because any descriptive account of "function" in the sciences will discard some occurrences of the concept "function" if they wholly contradict the account. But there are other axes along which the philosophical project of a theory of function can be considered. Some theories are conceptual analysis – and then, whether the analyzed concepts are ordinary functional statements (like in Wright 1973) or exclusively statements by biologists (e.g., Godfrey-Smith 1993) also makes a difference. Other theories are aiming at a theoretical redescription, in the context of a specific philosophical view of nature or mind – Millikan (1984) being the most famous example of such a strategy. Those various axes, along which one can situate the philosophical project about functional concepts and judgments, should be added to the general distinction made earlier, concerning what is taken to be the explanandum of a functional explanation (the presence of something or the dispositions of an encompassing system).

This should not lead either to relativism or to an attempt to decide which is the correct philosophical project. Each of them may have some legitimacy, but the important thing to keep in mind is that comparing two accounts of functions should be done on the consideration of their respective projects; differences between accounts are to be expected, if these accounts implement different philosophical projects. Some convergence in the end should be aimed at because an absolute discordance among the functional discourses and their interpretations would be very bad news for science, but the extent of such overlap is still undetermined.

A collective volume such as this one cannot therefore aim at providing the best up to date theories of function, except if all contributors were pursuing the same kind of project, which is not the case. Moreover, there is no attempt to discuss what, among the possible projects I outlined above, should be the best approach to functions and functional explanations, or the purpose of such investigation. The general assumption is that there is some legitimacy for preserving the plurality of approaches. But more precisely, even if all contributions vary concerning their commitment to a more normativist (e.g., Bouchard's chapter, or Walsh's) or to a more descriptivist approach in philosophy of science (e.g., Wimsatt's chapter, or Brandon's), there is a common idea that philosophical explorations about functions have to focus on – or, at least, be concerned with – actual scientific discourses and statements, in life sciences, or cognitive sciences. A philosophical view of functions which does not correspond in any way to such actual practice would indubitably fail, according to all contributors of this book. This may appear to be a very poor criterion of success for an account, but it highlights the fact that even if many chapters undertake conceptual analysis, such analysis may not be sufficient unless it is supplemented by an examination of the explanatory modalities along which the concepts are put to work in actual science and then connected to empirical data.

The multiplicity of projects undertaken in the same volume does not prevent it from answering general questions about functions and functional explanations. First off, the reader may get a sense that functions are used in such and such ways in, respectively, ecology (Chap. 5 by Bouchard), neuroscience of memory (Chap. 8 by Craver), or engineering (Chap. 11 by Vermaas and Houkes and Chap. 10 by Longy) from such a reading, and that it is hard to figure out a common account, even though some general features of the concept (e.g., its serving an explanatory role) appear. But a more elaborate reading will show that there are common issues, across these fields, with functions among which are the following ones: the univocity of the concept cannot be taken for granted; its metaphysical underpinnings along the lines of some functionalism (namely, the difference between functional and categorical states) are no more obvious; the scope of entities to which functions can be ascribed is not naturally defined and varies precisely according to our accounts of functions (here the chapter on functions of species by Bouchard (Chap. 5), as well as the section on function of oxygen molecules by Gayon (Chap. 4), is quite decisive); even if functional items may dysfunction, not all accounts of function justifies that robust claims of abnormality can be established; identifying mechanisms in systems yields one sense of functional statements but there may be a more ontologically consistent notion of function than this one, which is dependent upon the systems one defines. For this latter reason, and especially because philosophers in general may worry whether functions and functional properties are part of the furniture of the world (exactly as metaphysicians worried about dispositions, and interestingly Mumford (1998) answered this question by defining dispositions in functional terms ...), the relevance of the present investigations for philosophers in general is not a fiction.

The volume contains two parts; a section on "biological functions and functional explanations: genes, cells, organisms, and ecosystems," and another on "psychology, philosophy of mind, and technology: functions in a man's world." To some extent this division parallels a dichotomy that one could find in the development of the debates about functions and functional explanations. These debates have been vividly fueled by both issues about biology and issues about psychology. Wright's account, mentioning natural selection, was quickly taken up by philosophers of biology, whereas Cummins' account was first intended to make sense of classical explanatory schemes in psychology. Therefore, psychology and biology were, from the beginning, differently positioned regarding the two main rival views of function. Clearly, concerning biology, it could even be argued that the interest of philosophers in general, especially philosophers of action such as Wright, met the interest of biologists concerning the role of natural selection. Some of the first papers about the etiological theory were indeed written by precisely the first generation of academic philosophers of biology (Rosenberg, Ruse, etc.) and often by prominent evolutionary biologists such as Ayala. Remember that in the early days of the philosophy of biology, Mayr, possibly the most influential biologist (for philosophers of biology), explained that what is really proper to biology is evolution by natural selection (not physiology, which is chemistry, etc. Mayr 1961). Therefore, a view of function that centers on natural selection was easily accepted as biologically adequate by most philosophers of biology. Inversely, given the prevalence of the cognitive classical paradigm in psychology in those days, and the analogy with computers, a view of functions akin to the function of computer modules was easy to embrace by philosophers of psychology.

Hence, the first section will investigate theories of function and functional explanations in the light of what is going on currently in the life sciences, especially the issues listed above. In particular, some kinds of biological entities claim the attention of philosophers regarding the functions ascribed to them, either because it is hard to think that they are undergoing selection or because they make room for another level of selection besides the one which is classically accepted (e.g., ecosystems, inorganic molecules …). Biology – from molecular biology to ecology – concerns entities of various size; some of them may not be wholly biological but still, crucially, interact with the biological domain and therefore have their place in some life sciences. Gas molecules in the body, as well as ecosystems, are at least partially abiotic and are however crucial to living beings, but our current theories of function may not account for cases where such very large or small entities are ascribed functions, because they are tailored to suit the functional ascriptions to more conventionally live beings (organs, behaviors, cells, proteins …). Hence this section often considers various scales in biological functional statements. It is also concerned with the general organization of living systems and how coarse and fine grained functional ascriptions of multiple kinds may be stated.

Given the variety of accounts of function, and the often repeated claim that no single account can capture both of the legitimate uses of the concept, one main issue will be first making sense of and then assessing pluralism. This constitutes the second part of the section, and given the wide acceptance of etiological theories among philosophers of biology, it will raise problems specific to the dominant formulation of the etiological theory.

The second section, about mind, psychology, and techniques, reflects the dual orientation of philosophers at the beginning of the debates. From the beginning of the controversies, there has been a crucial topic: whether functions are ascribed in the same sense to biological creatures and to human artifacts and social structures – which was Wright's original position – or whether there is an irreducible difference between both. Even if an account of functions in biology (be it dogmatic, pluralist etc.) is found, there is still the issue of the possible extension of an etiological account of functions to men's institutions and artifacts, given that natural selection is not so pervasive and efficacious in human history at the first glance – whereas no principled problem has affected a causal-role view of functions. This second section deals with the concept of function in areas where human choice, selection, and intention at least make room for functions (Mc Laughlin 2001) by endowing a state of affairs with practical meaning for human plans.

The first part of the second section concerns the metaphysics of functions and the connection between psychological or biological functions and functionalism – given that, metaphysically, it is plausible that both kinds of function raise common problems, for which one should avoid some common misconceptions. The last section concerns an area where functional talk is crucial, and perhaps its almost original locus, which are artifact, technique, and design. Here, one deals with the ontology of things made and endowed with functions and goals related by definition to their creators, and even defined or potentially defined by them. The question of the ontological underpinnings of the concepts of functions has clearly to be raised in this domain. Peter Mc Laughlin (2001) argued that if the epistemology of functional ascriptions, which may be similar regarding organisms and artifacts, should be supplemented by an ontological perspective on functional items. To this extent, he argued that the ontology of biological functions requires beings which genuinely reproduce, whereas the ontology of artifacts do not require something more than particulars, states of affairs, and propositional attitudes, given that an artifact is a state of affair X endowed with the intention of making Y through X (i.e., a propositional attitude). Raising the problem of the commonality between artifact and biological functions thereby forces one to consider the ontological underpinnings of functional discourse.

3 Contributions in Detail

The first part of the first section considers the functional discourse in its relation to the development and organization of living systems. William Wimsatt, provided one of the first detailed analyses of functional explanations (Wimsatt 1972), identifying an etiological-selective as well as a theory-laden systemic perspectives. His current contribution addresses the issue of the architecture of organisms. It seems that functional traits in an organism cannot yield the systematic structure of organisms by themselves because each of them fulfills its function but such fulfilling does not ipso facto entail a systematic connection with other functional traits. So there is the issue of understanding how the functional architecture, which scales across several levels (genes and their expression networks, cells, organs, systems etc.) is articulated with the set of independent functional traits identified by a functional analysis. Wimsatt's sophisticated theory acknowledges networks of conditioning at several levels between traits as nodes and showed how the depiction of such treelike architecture allows functional ascriptions and explanations. He highlights the key role of robustness at all levels in such architecture. Such a contribution helps to disentangle functional explanations from a purely functionalist or adaptationist view of organisms, which has been under attack for three decades now, starting with Gould and Lewontin's famous paper on adaptationism (1978). More precisely, given the increasing concern with architecture with all levels of living systems (i.e., the architecture of the genome, of the nervous system, the cell, or of the brain, and the correlated attention to the role of networks such as gene regulation networks or cell metabolism networks, with their properties of robustness, redundancy, etc.), a concern which somehow downplays the explanatory role of selection and adaptation, Wimsatt's paper provides a renewed and detailed understanding of function that suggests how functional perspectives are still grounded and relevant in such a context, and how they are carried on by researchers.

The contribution of Denis Walsh (Chap. 3) also concerns developments in recent evolutionary biology which call for "extending the evolutionary synthesis" on the basis of a new understanding of development and its role in evolution, highlighted above (C2). The alternative views advocated by evolutionary developmental biology theorists in general displace the center of gravity of evolutionary biology from genes to the developmental potentialities of organism. Variation not only is the mutation and recombination of genes but also relies on the active restructuring of developmental modules and toolkits, which accounts, in part, for the main evolutionary novelties (e.g., Muller and Newman 2005). So, the potential for variation accounted for in these developmental terms is at least as relevant as natural selection (which acts upon these variations), with regard to the evolution of forms and behaviors. In this context, the functions of traits of organisms cannot be solely understood in etiological terms with relation to natural selection. Walsh will anchor a new understanding of these functions in the theories of adaptive active responses of organisms to environmental change, as investigated by West-Eberhardt (2003). This theory offers a radical way of answering the challenges that the new Evo-Devo theory, and "extended synthesis" proponents (e.g. Pigliucci and Müller 2011), present to the etiological theories of functions (above, C1), which have been elaborated in the framework of classical, modern-synthesis style, evolutionary theory. Even though the contribution of natural selection to the explanation of traits can be undermined, traits, organs and behaviors can still legitimately be ascribed functions without having to appeal to causal-role functions, whose drawback is that they can hardly be taken in a realist way (i.e., as properties really existing in nature).

The chapter (Chap. 4) by Jean Gayon also questions the etiological theory of functions widely admitted by philosophers interested in evolutionary biology. Gayon's question bears upon the range of entities likely to have a function in etiological terms; no specification of what can have a function is given by etiological theorists, especially because selection is likely to act upon organs, but also behaviors, or traits like sex ratio, so that the inclusion within organism, or the material or structural composition, is orthogonal to whether a trait has a function and which one. Gayon notes that there is a discrepancy between the theory and the kinds of common statements they make about some entities such as oxygen. If oxygen has a function, as physiologists continuously say it does, it must have been selected for etiological theorists; who find it hard to admit. Therefore, the etiological theory of function either needs a radical reshaping or is at odds with a significant part of the biologists' use of it.

The second part of the first section addresses the etiological theory more directly and revolves the issues around pluralism and realism. Evolutionary biology and ecology, rather than development and physiology, are under focus. The chapter

(Chap. 5) by Frederic Bouchard is a plea for using causal-role functions in ecology. Amundson and Lauder put forward a first and famous defense of causal-role accounts of function (1994), considering the case of functional morphology. Here, building on very recent literature in ecology which considers a possible community selection, Bouchard urges us to see functions in ecology as plausibly understandable in causal-role terms. This view is tied to a revisionist conception of fitness, which detaches it from replication and keeps the mere component of persistence, now ascribed to lineages (and not organisms), in order to answer some challenges faced by traditional views of fitness and selection. The issues raised by philosophers about fitness and the nature of selection (above, C1) are therefore reflected in this contribution, which takes controversial studies about high level selection at face value.

Robert Brandon replies to this argument with a defense of pluralism in evolutionary biology and ecology. His approach distinguishes historical and unhistorical views of function, respectively, considering the causal-role view, advocated by Bouchard, as an unhistorical conception, as well as theories which see functions as contributions to current fitness in the same way as behavioral ecologists often consider adaptations to be highest fitness traits without considering history (e.g., Reeve and Sherman 1993). Then he shows the complementarity of these perspectives, using a parallel with geological concepts of mountain. Thereby, in ecology as well as in population biology, both accounts of function define two equally legitimate approaches to the issue of functional traits, but with different conditions of validity. Pragmatism related to which conditions are best for using one or the other concept, go together with the pluralism regarding function concepts. Now that we are moving towards a higher integration of ecological sciences and evolutionary biology, the pluralism of function concepts may here be adequate to address the variety of functional explanations that are often used at the same time in various areas of ecology: eco-systemic functions, functional equivalence between species, and function of traits in behavioral ecology.

Huneman's chapter (Chap. 7) also advocates pluralism, but based on his treatment of the issue of justifying fine grained functional ascriptions in the framework of etiological theories. Distinguishing functional ascriptions and functional explanations, he claims that in order to disambiguate various candidate functions for traits, one should pick up a specific explanatory strategy within which to embed the functional ascriptions. Therefore, given that something in this choice of strategy pertains to the sole explanatory interests, the realism of etiological theory has to be weakened to make room for such explanatory dependence.

Taken as a whole, this part of section 1 provides a systematic view of the reasons for embracing some pluralism when trying to make sense of functional concepts and explanations in current evolutionary biology.

The first part of the second section extensively considers the use of functional concepts in cognitive sciences and philosophy of mind. Carl Craver (Chap. 8) considers the nested architecture of cognitive systems made of mechanisms packed in higher level mechanisms. In this context, he explains how causal-role functions can be used to answer questions about the causal structure of a mechanism, but, building on the examination of the case of ion – channel in the neuroscience of memory, he also provides a finer view of the richness of functional concepts in this science. The chapter finally distinguishes three perspectives: causal, constitutive and etiological, from which one can legitimately ask questions about a cognitive system. Pragmatic considerations are therefore required to discriminate between varieties of functional concepts.

Then Carl Gillett pulls the topic of functions in psychology and cognitive sciences to within the general frame of an enquiry about what functional properties should be from a metaphysical point of view. He particularly contrasts the view of functionalism tied to the logical machinery of Ramsey sentences, used, for example, by David Lewis, and the functionalism which would be built on an examination of the making of empirical science in psychology. This difference then leads to a reassessment of the notions of realization that are involved in defining functional properties, given that being functional properties, in general, presupposes some multirealizability. Therefore, Gillett's chapter (Chap. 9) links the philosophical discussion of functions in the science to the revival of metaphysical questions about realizers and realization, initiated in the 2000s, highlighted above. Realization and realizers are defined in relation to causal properties and interactions, and to this extent, the chapter raises issues about the links between function and causation. Though this chapter does not deal directly with one of the usual family of conceptions of functions, namely, etiological or systemic ones, it exemplifies the ways in which "metaphysics of science", that is, metaphysics informed by science, addresses the same issues as the ones involved in the debates about functions.

Taken together, this part of the second section provides a much richer picture of the possible connections between causation and functions in the field of psychology and cognitive science, tied closely to an investigation into explanatory strategies in the field.

The second part of the section considers another aspect of the use of functional concepts regarding human existence and activities, that is, artifacts and technology. A general issue, present from Wright 1973 paper, is the possibility of having a unified account of functional concepts in life sciences and techniques. Biologists extensively use what Lewens (2004) called the artifact scheme, namely, considering organismal traits as parts of machines and enquiring about their function in the same way as someone who found an unknown machine would investigate its various functions. Functional terms are used indifferently to describe machines and artifacts. However, if one wants a rigorous unified account, many difficulties emerge – for example, the fact that in the analogy of organisms and artifacts, organisms themselves do not have functions, only their parts do, unlike artifacts which have functions as a whole. Offering a unified account would also mean superseding the intuitive idea that functions of artifacts are intentions of their designers or users, whereas in biology, intention has no legitimacy.

Vermaas and Houkes' chapter (Chap. 11) present a synthetic theory which, at the same time, acknowledges that the etiological theory of functions account for many of the biological uses of functions and some aspects of the artifact functions and includes an irreducible intentional component when it comes to artifact. In contrast,

Longy's chapter (Chap. 10) tries to present a wholly unified account of functions, which gets rid of intention in the definition of the nature of artifacts, solving the issue (pointed out by Vermaas and Houkes) of the first invented token of an artifact. For Longy, subjective and objective properties are not exactly distinguished by some relation to the human subject; therefore, she defends the idea that, even if artifacts are created, there is something objective in their having the functions that they have. To this extent, they can be subsumed under a theory which makes sense of functional properties as existing objectively, and precisely this is done by the etiological theory in the sense that it traces functional notions back to causal history (which is objective). Therefore, a generalized selection history is likely to account for functional ascriptions in general, be they in the field of biology or in the domain of man-made (and used, and exchanged …) artifacts.

Finally, the book ends with an epilogue written by Larry Wright, whose 1973 paper on functions largely contributed to initiate rich philosophical conversations about functions by providing the first and clearest expression of the etiological theory. Wright's chapter puts the whole debate, and especially the set of contributions that follows, into a new light because he offers a personal perspective on the series of debates and advances that followed his paper and formed the context of the present book. In particular, whereas all of the contributions are deeply entrenched within philosophy of science, Wright shows the links between general issues in rationality and action theory and these considerations. Extending the notion of function into an idea of teleology, he indicates ways of making sense of the pluralism regarding (scientific) uses of functional terms as inscribed within the general frame of action and rationality.

References

Abrams, M. 2009. The unity of fitness. *Philosophy of Science, Philosophy of Science* 76: 750–761.

Amundson, R. 2005. *The changing role of the embryo in evolution*. Cambridge: Cambridge University Press.

Amundson, R., and G.V. Lauder. 1994. Function without purpose: The uses of causal role function in evolutionary biology. *Biology and Philosophy* 9: 443–470.

Bechtel, W., and A.A. Abrahamsen. 2005. Explanation: A mechanistic alternative. *Studies in History and Philosophy of Biology and Biomedical Sciences* 36: 421–441.

Bechtel, W., and R. Richardson. 1993. *Discovering complexity*. Princeton: Princeton University Press.

Bekoff, M., C. Allen, and G. Lauder (eds.). 1998. *Nature's purposes*. Cambridge: MIT Press.

Bigelow, R., and R. Pargetter. 1987. Functions. *The Journal of Philosophy* 84: 181–197.

Bouchard, F. 2009. Causal processes, fitness and the differential persistence of lineages. Philosophy of Science 75(2009): 560–570.

Brandon, R. 2010. Natural selection. *The Stanford Encyclopedia of Philosophy* (Fall 2010 Edition). Edward N. Zalta (ed.). http://plato.stanford.edu/archives/fall2010/entries/natural-selection/.

Block, Ned. 1997. Anti-reductionism slaps back. In *Philosophical perspectives 11: Mind, causation, and world*, ed. J.E. Tomberlin, 107–132. Boston: Blackwell.

Buller, D. 1998. Etiological theories of function: A geographical survey. *Biology and Philosophy* 13: 505–527.

Buller, D. (ed.). 1999. *Function, selection and design*. Albany: SUNY Press.
Burt, A., and R. Trivers. 2006. *Genes in conflict*. New Haven: Harvard University Press.
Craver, C.F. 2001. Role functions, mechanism, and hierarchy. *Philosophy of Science* 68: 53–74.
Craver, C. 2008. *Explaining the brain: Mechanisms and the mosaic unity of neuroscience*. New York: Oxford University Press.
Cummins, R. 1975. Functional analysis. *The Journal of Philosophy* 72: 741–764.
Damuth, J., and I.L. Heisler. 1988. 'Alternative formulations of multi-level selection', selection in structured populations. *Biology and Philosophy* 17(4): 477–517.
Darden, L. 2006. *Reasoning in biological discoveries*. Cambridge: Cambridge University Press.
Enç, B., and F. Adams. 1992. Functions and goal directedness. *Philosophy of Science* 59(4): 635–654.
Fodor, J. 1974. Special sciences: Or the disunity of science as a working hypothesis. *Synthese* 28: 97–115.
Gillett, C. 2003. The metaphysics of realization, multiple realizability and the special sciences. *The Journal of Philosophy* 100: 591–603.
Gilbert S. 2003. Developmental biology. Boston: Sinauer.
Ginzburg, L., and M. Colyvan. 2004. *Ecological orbits: How planets move and populations grow*. New York: Oxford University Press.
Glennan, S.S. 1996. Mechanisms and the nature of causation. *Erkenntnis* 44: 49–71.
Godfrey-Smith, P. 1993. Functions: Consensus without unity. *Pacific Philosophical Quarterly* 74: 196–208.
Godfrey-Smith, P. 2009. *Darwinian populations and natural selection*. Oxford: Oxford University Press.
Gould, S.J., and R.C. Lewontin. 1978. The spandrels of San Marco and the Panglossian paradigm: A critique of the adaptationist programme. *Proceedings of the Royal Society of London, Series B: Biological Sciences* 205: 581–598.
Griesemer, J. 2000. The units of evolutionary transitions. *Selection* 1: 67–80.
Hubbell, S.P. 2001. The unified neutral theory of biodiversity and biogeography. Princeton: Princeton University Press.
Huneman, P. 2012. Natural selection: A case for the counterfactual approach. *Erkenntnis* 76(2): 171–194.
Huneman, P. 2013. Function and adaptation: Conceptual demarcation. In *Functions and functional explanations in biomedical and human sciences*, ed. J. Gayon and A. de Ricqlès. Dordrecht: Springer.
Kingsland, S.E. 1985. *Modeling nature: Episodes in the history of population ecology*. Chicago: University of Chicago Press.
Kitcher, P. 1993. Function and design. *Philosophy of Science* 58: 168–184.
Leibold, M.A., M. Holyoak, N. Moquet, et al. 2004. The metacommunity concept: A framework for multi-scale community ecology. *Ecology Letters* 7: 601–613.
Lewens, T. 2004. Organisms and artifacts: Design in nature and elsewhere. Cambridge, MA: MIT Press.
Lewens, T. 2009. The natures of selection. The British Journal for the Philosophy of Science 61(2): 1–21.
Lewontin, R. C. 1970. The Units of Selection. Annual Reviews of Ecology and Systematics 1: 1–18.
Machamer, P.K., L. Darden, and C.F. Craver. 2000. Thinking about mechanisms. *Philosophy of Science* 67: 1–25.
Matthen, M., and A. Ariew. 2009. Selection and causation. *Philosophy of Science* 76: 201–224.
Mayr, E. 1961. Cause and effect in biology. *Science* 134: 1501–1506.
McLaughlin, P. 2001. *What functions explain*. Cambridge: Cambridge University Press.
Michod, R. 1999. *Darwinian dynamics*. New York: Oxford University Press.
Millikan, R. 1984. *Language, thought, and other biological categories: New foundations for realism*. Cambridge, MA: MIT Press.

Millikan, R. 2001. Biosemantics. *The Journal of Philosophy* 86: 281–297.
Müller G., Newman S. 2005. J. Exp. Zoo. 304 B 6: 487-503.
Mumford S. 1998. Dispositions. Oxford: Clarendon Press.
Neander, K. 1988. What does natural selection explain? Correction to Sober, Phi-losophy of Science 55(3): 422–426.
Neander, K. 1991. Functions as selected effects: The conceptual analysts defense. *Philosophy of Science* 58: 168–184.
Nilsson, D.E., and S. Pelger. 1994. A pessimistic estimate of the time required for an eye to evolve. *Proceedings of the Royal Society of London* B 22, 256, 1345: 53–58.
Odling-Smee, J., K. Laland, and M. Feldman. 2003. *Niche construction, the neglected process in evolution*. Princeton: Princeton University Press.
Okasha, S. 2006. *The levels of selection in evolution*. Cambridge: Cambridge University Press.
Pigliucci, M. 2007. Do we need an extended evolutionary synthesis? *Evolution* 61(12): 2743–2749.
Pigliucci, M., and G. Müller. 2011. *The extended synthesis*. Cambridge: MIT Press.
Polger, T. 2004. *Natural minds*. Cambridge: MIT Press.
Polger, T. 2007. Realization and the metaphysics of mind. *Australasian Journal of Philosophy* 85(2): 233–259.
Putnam, H. 1967. Psychological predicates. In *Art, mind, and religion*, ed. W.H. Capitan and D.D. Merrill, 37–48. Pittsburgh: University of Pittsburgh Press.
Raff, R. 1996. *The shape of life: Genes, development, and the evolution of animal form*. Chicago: University of Chicago Press.
Reeve, H.K., and P.W. Sherman. 1993. Adaptation and the goals of evolutionary research. The Quarterly Review of Biology 68: 1–32.
Shapiro, L. 2010. *Embedded cognition*. London: Routledge.
Shapiro, L. 2000. Multiple realizations. *The Journal of Philosophy* 97: 635–654.
Sober, E. 1995. Natural selection and distributive explanation: A reply to Neander, The British journal for the philosophy of science 46(3): 384–397.
Sober, E., and D.S. Wilson. 1998. *Unto others*. New Haven: Harvard University Press.
Wright, L. 1973. Functions. *Philosophical Review* 85: 70–86.
Walsh, D., T. Lewens, and A. Ariew. 2002. Trials of life: Natural selection and randomdrift. *Philosophy of Science* 69: 452–473.
West-Eberhardt, M.J. 2003. *Developmental plasticity and evolution*. New York: Oxford University Press.
Wimsatt, W. 1972. Teleology and the logical structure of function statements. *Studies in History and Philosophy of Science* 3: 1–80.

Part I
Biological Functions and Functional Explanations: Genes, Cells, Organisms and Ecosystems – Functions, Organization and Development in Life Sciences

Evolution and the Stability of Functional Architectures

William C. Wimsatt

Abstract A puzzle about functional organization has gone largely unnoticed in our debates about the nature of function. Although we recognize that systems evolve and acquire new functions, noone has systematically discussed whether there are constraints on how this is likely to occur. This naturally suggests the widely discussed problem of evolutionary innovation, but I am interested here in the complementary problem: things we already know about epistatic codependencies in functional organization suggest that the common conservation of organization under mutations or sexual recombination should be quite remarkable. This arises either for Cummins style role function or on the selectionist account of function. There are strong constraints on the addition of new functions imposed by conditions of evolvability and generative entrenchment. Evolvability favors increases in robustness for functionally important architectural features. And greater generative entrenchment produces more constraints on what changes can be adaptive, yielding a rapidly declining probability that macromutations of increasing size will work. These are both bound to affect the structure of functional architectures and the character of functional innovations. If these constraints were violated systematically and frequently, the flux of changes in functional architectures would give us significant troubles even in individuating and identifying functions in complex organizations and the instability of function would make evolution virtually impossible. But taking account of these constraints is more broadly revealing. Doing so gives a new perspective on the relation between selectionist and causal role accounts of function.

W.C. Wimsatt (✉)
Department of Philosophy, Committees on Evolutionary Biology and Conceptual and Historical Studies of Science, The University of Chicago, Chicago, IL, USA
e-mail: wwim@uchicago.edu

P. Huneman (ed.), *Functions: selection and mechanisms*, Synthese Library, Studies in Epistemology, Logic, Methodology, and Philosophy of Science 363, DOI 10.1007/978-94-007-5304-4_2,

1 A Concept of Function

In 1972 I proposed an analysis of function that argued for a common underlying structure for all cases of functional attribution, whether of human plans, intentions, or artifacts, or of biological organisms, or of parts of any adaptively organized evolving system, whether a product of conscious intention or of natural selection (Wimsatt 1972). Complex functional organization, whether a natural product, a product of human intentional action, or of cumulative cultural change, reflects the action of selection processes acting at one or more levels of organization.

In referring to selection, this analysis is similar to those advanced a year later by Wright (1973) or by Millikan in 1984. But whereas Wright and Millikan made a past history of selection a requirement for the existence of a function, I sought to relate function to the ongoing dynamics of selection, like that later advanced by Bigelow and Pargetter (1987).[1] Thus, I related what it was for a behavior or action to be functional to whether it would have a positive effect on probability of survival of the relevant evolutionary unit (a fitness measure) over what had gone before, or, for human intentional action, whether it increased the probability of attaining the purpose of the action. I wanted to be able to discuss whether a mutation or change in behavior or plan was functional (positive), neutral, or dysfunctional (negative) in selective effect when it first occurs, since these marginal effects on fitness would presumably be instrumental in whether it was incorporated or lost. This was the knife edge of selection as it formed and elaborated adaptation. I also wanted to be able to evaluate function in the context of optimization arguments that could identify the presumed direction of selection.[2] Each of these reflects directly how and why function talk was an essential part of the apparatus for applying a selectionist theory. To give Wright (and later Millikan) their due, it seems plausible that one might not speak of *having* a function (Wright's target of analysis) until that utility

[1] Bigelow and Pargetter note the similarities between our analyses but also suggest differences that do not exist: (1) As they suggest, my analysis is in terms of probabilities, which they assume are to be interpreted in a frequentist manner. But I made it clear that they are to be taken as supporting counterfactuals and supported by underlying mechanisms (*Ceteris paribus* qualified to allow for individual variations), which make them functionally equivalent to propensities. (The explicit use of propensities in such contexts dates from Mills and Beatty in 1979.) (2) It is also misleading to describe my theory as a "goal-oriented" account (their footnote 1). Goal-directed behavior is (with some qualifications) a subclass of functional behavior and is explicitly criticized as an inadequate basis for a general analysis of function on pp. 20–22 of my (1972). (3) There are many explicit discussions of problems in inferring function that arise through the choice of a reference situation not having that function for comparison (1972, pp. 55–61) that could only be satisfied for a selectionist account of functional inference and which presage my concerns here.

[2] These are not entirely independent of the selection history, since, for example, the configuration of constraints determining the form of an optimization is a product both of the existing functional architecture—a product of selection history—and of the structure of environmental variation. But current functionality of a mutation does not require that this history be available and subject to evaluation—only that selection currently favors it.

had been demonstrated through the actions of selection in prior generations, but pursuing *current functionality* as the primary target of analysis, and then looking at the historical consequences of realizing a new function in this way also allows one to escape many counterexamples that arise from the requirement of an etiological history as a *conceptual* requirement for function.

Nonetheless, being functional or having a function of any complexity makes an etiological history overwhelmingly likely, as an *empirical* matter—and the more complex the function, the more extended and complex a historical trajectory we may expect. The history of current functions should characteristically be a sequence of co-opted earlier structures and exaptive features that undergo further functional and structural changes (Gould and Vrba 1982). But quite a bit more can be said. I analyzed some of the conditions on such functional inferences in 2002. Extensions of the same conditions that make an etiological history likely for a current functional relationship also make it likely that at least the deeper features of any existing functional architecture will have greater antiquity, generality, adaptive, and even conceptual necessity because their causal depth and entrenchment will make them more stable than more superficial features. I consider this at length below since it bears directly on the main focus of this chapter.

A second difference between Wright's analysis and mine is more cosmetic or tactical than fundamental. I agree with Wright that selection is centrally involved with function, but I pursued that general analysis of function (1972, 2002) in terms of purposes rather than explicitly putting selection in the analysis for three reasons:

1. Purposes *in general* have a deep connection with selection and could plausibly be thought *always* to involve and be unpacked in terms of selection processes—even as a conceptual matter. Changing to selectionist language only affects vocabulary. (It does thus not, e.g., make it nonteleological.) To emphasize that my aim was not to "translate away" talk of purposes in terms of selection, I left the analysis in terms of purposes and made the point about selection elsewhere in the analysis.
2. Talk of purposes themselves has a rich logic that I wanted to exhibit also for those who resisted the selectionist analysis, while at the same time indicating how far a selectionist analysis could reach through capturing all of the logical features normally associated with purposiveness.[3] But this means also that this "consciousness-free" notion of purpose construed in a way suggested by the analysis of function in biology could play a central (though not exhaustive or eliminative) role in the analysis of human purposiveness and intentionality. These latter notions are multi-layered, and anything that we can do to peel away, uncontroversially, important elements of their structure is worthwhile. I think that consciousness has nothing to do with many and perhaps all important features of purposiveness per se.

[3] I do think that understanding the power of the selectionist analysis, in context, removes most of the resistance to regarding it as an analysis of the logic of human purposeful behavior—while at the same time recognizing that there are many features of human psychology and language use that it does not capture.

3. I wanted to make the point that a class of theories of a certain form, selectionist theories, could justify (and more: require) talk of purposiveness in contexts where there were no conscious intentions, indeed, no consciousness involved. These theories generally seek to explain features of functionally organized systems in terms of cumulative evolutionary processes. In these theories purposes would appear as a kind of theoretical construct (Wimsatt 1972).

2 A General Form for Attributions of Function and Some of Its Consequences

The fullest and most paradigmatic functional attributions found in either evolutionary contexts or in human purposive contexts have a structure I elaborate here. (See also Figure 1, from my work (1972).) Functional attributions do not always mention all of these elements, which may in some contexts be taken for granted, but they are all logical elements of any such attribution that follow from the normal expectations surrounding such statements. In 1972 I provide separate analyses for the selectionist, evaluative, and causal role interpretations that treat these as special limiting cases of the general account. This thus covers and demonstrates many fundamental similarities and relationships between the selectionist and the causal role accounts. A similar approach on these points is taken by Barker (2008).

I elaborate this structure through the use of a *normal form* for function statements as follows:

According to causal theories $\{\mathbf{T}_j\}$ and relative to purpose or selection criteria $\{\mathbf{P}_k\}$, *the* function of behavior **B** of item ***i*** in system **S** in environment **E** is to do or to bring about causal consequence **C**.

The variables in this statement refer to all of the elements necessary for making a function statement, though not all of the information necessary for assessing its truth.

This second task must also include states of that and other systems necessary for making the comparative assessments to determine whether the functional item contributed to purpose attainment, and other information about the system (including non- and dysfunctional interactions) necessary to evaluate its net contribution. It is called a "normal form" because many attributions of function do not make all of the relevant variables explicit in the attribution.

Any element serves its function in terms of what it does in some larger system and will characteristically do so by producing some consequence of its operation or presence in some environment or set of environments. (If it does so in virtually all environments, or the environment is understood, it may not be mentioned.)[4] The functional behavior does so according to the mechanisms and interactions involved according to a set of relevant causal theories. This much is shared by both

[4] This is plausibly responsible for claims that there are notions of function that make no reference to the environment. This claim (by Lauder and Amundson 1994) is discussed further below.

teleological (selectionist) and nonteleological senses of function. In any selective or evaluative context, its functional performance is evaluated according to another set of causal theories and mechanisms concerning how it contributes to the attainment of some state or set of states—its purpose. These theories are presumed to specify what it is about its operation that is selected. In causal role function, this last set of theories that provides standards or ends for evaluating performance is either missing or not invoked. This may be (e.g., in medicine or in functional morphology) because the standards are taken for granted.

There are many qualifications and elaborations required here because the conditions for such a comparative evaluation are often complex and counterfactual, and the necessary information may be lacking, especially for evaluating the more deeply embedded and entrenched components. In these contexts, a role-function evaluation may be used or even required simply because there is no information of the sort required for a comparative evaluation: any actual comparative case is too phylogenetically and functionally different to permit localizing praise and blame among the diverse differences in attributing function (See Wimsatt 1972, 55–59.) This undoubtedly helps to explain why causal role-functional analyses are so common in areas like functional morphology, where one is dealing with idealized archetypes rather than intraspecific variation, and the features under discussion *have no variation in the relevant respects*. Nonetheless, it is ultimately selection that is the driver in creating these complex organizational features and divergent functional complexes. Cummins' style role functions can be picked out but are relatively uninteresting and ad hoc when there is no evolutionary or selectionist or evaluative process operating in the background. (These are the imagined counterexamples of mechanisms created ex nihilo by statistical mechanical accidents.) These are degenerate cases and should be ignored on either account but can be more easily dismissed on selectionist ones.[5] The fact that a unique function is (provisionally) claimed relative to these variables allows this schematic normal form to assume the form of a mathematical function[6]:

$$\mathbf{F}\left[\mathbf{B}(i),\mathbf{S},\mathbf{E},\mathbf{P},\mathbf{T}\right]=\textbf{to do C}.$$

The "item" variable appears in lower case because it is an index variable, not independent from the behavior variable, which will always be behavior of some object, **B(*i*)**. This has another important implication. Changing ***i*** as long as it preserves the behavior (and the context provided by the other variables) under consideration

[5] Though even this is no guarantee for selection needs to be marked off from differential stability. In Wimsatt (1972), pp. 16–17, I note that selection can shade into stability as the systems get simpler, an insight later used to good advantage by Dawkins (1976) in his reductionistic account of genic selection. Even feedback can be difficult to distinguish from simpler processes (e.g., steady-state equilibration in an open system) in simpler systems (Wimsatt 1971).

[6] The main function of the uniqueness claim is heuristic and comes closest to being satisfied for relatively modular systems (Wimsatt 1972). As parts of a system become more functionally interdependent, there are more situations where "closed functional loops" are formed, and the assumption can be expected to be violated more frequently.

makes no difference in the equation and shouldn't. It would make no functional difference and yields what is called a "functional equivalent." (An object has a function in virtue of what it *does*, not in virtue of what it *is*.) Functional equivalents are items that behave in the same way in relevant respects (for accomplishing the function) under the same circumstances. This redundancy of the "item" variable permits an unusually systematic treatment of functional equivalence and in addition provides a matrix for the functional organization that permits and encourages a systematic classification and analysis of any functional differences that *do* occur. There have been several interesting discussions recently of the notion of functional equivalence and the related notion of multiple realizability. I return to this in Sect. 5.

The form of the function statement schema also naturally suggests the logical form of functional organization (Wimsatt 2002), in which behaviors of items have possibly multiple distinct consequences that serve functions in distinct systems and or different environments, according to the different criteria of evaluative purposes (e.g., viability and reproduction of organisms, groups, and any other entities that meet the conditions to be units of selection (Wimsatt 1980, 1981b; Lloyd 1988; Okasha 2006)). Philosophers have tended to focus on cases where multiple possible alternative (but competing) and usually imagined purposes are considered (e.g., the pumping the blood versus heart sounds for heart behaviors). The situation where there are multiple units of selection at different levels of organization suggests that an entity might frequently serve parallel different but similar sounding functions (e.g., survival of the evolutionary unit) for several distinct units (e.g., gamete, organism, and population in Lewontin and Dunn's famous case of the t-allele in mice, 1960) or serve a function for some but not for others of the units. Having a schema with all of these as free variables encourages the exploration of the multiplicity of functional (and dysfunctional) connections that an entity might have with other interacting systems and subsystems, and this was part of the motivation for seeking such an abstract schema.

3 Small Mutations as the Raw Material for Changes in Functional Organization

The focus of this chapter is how functional organization reacts to changes, but we haven't yet talked about the character of those changes. Quasi-continuous slow selection and uniformitarian incrementalism has been a deep assumption of evolutionary theory since Darwin's time. A similar perspective is urged by Basalla (1987) for technological change and, in different ways, is supposed by Simon (1969/1996) and Campbell (1974) for cultural change. Most of Darwin's contemporaries expressed strong doubts (as Huxley did) and espoused a more saltative view of evolution. A number of changed perspectives have pushed opinion further in that direction since. This includes:

1. The "punctuated equilibrium" theory of evolutionary change espoused by Eldridge and Gould in 1972 and other venues since.

2. Evidence of catastrophic perturbations of the environment due to bolide collisions, volcanic eruptions and massive lava flows, and massive carbon dioxide and other driven temperature fluctuations leading to global freezes and heat deaths, and then leading to mass extinctions (Alvarez et al. 1980; Raup 1993).
3. The work of the Grants and their many students in the Galapagos (Grant and Grant 2008; Price 2008) and others demonstrating that selection could be far more intense and evolutionary change far more rapid than supposed.
4. The rise of evolutionary developmental biology, which has directed greater attention to macroevolution, an appreciation of the importance of rare events and led to more dissatisfaction with last generation's assumptions that macroevolutionary processes, were adequately characterized as a simple extrapolation of observed microevolutionary processes.

Nonetheless, arguments suggest that smaller adaptive changes are both far more frequent than larger ones and also more frequently adaptive than supposed. The rise of catastrophist scenarios has not made these less relevant. The first of these proceeds from a probabilistic model of the increasing number of features of the phenotype affected by mutations of increasing size (e.g., Wimsatt 1986; Schank and Wimsatt 1988).[7] If each feature can be affected positively or negatively and can also affect other features, a growth in the number of things affected will be expected to have an increasingly negative effect. If each feature *should* be changed in a manner tuned to the others for adaptive effect, the probability of net adaptive change would be calculated according to the multiplicative law for independent events and would decline exponentially. So larger mutations should become exponentially more likely to lead to catastrophic failures. Early acting mutations do commonly have disastrous consequences.[8] Thus, earlier stages of development would tend statistically to be more evolutionarily conservative, basically because their features were used for more things downstream and tend to cause cascading disruptions in development. This process, wherever it occurs, I call "generative entrenchment" (Wimsatt 1986). The presence of differential generative entrenchment (different degrees of downstream dependency for different elements in a complex system) is robust and generic—essentially unavoidable (Wimsatt and Schank 1988, 2004; Wimsatt 2001). This bias toward smaller adaptive mutants means that gradualism is dominant, though not universal. These arguments apply also for technology, where larger adaptive change is facilitated in various ways, though still difficult and much rarer than smaller ones (Wimsatt and Griesemer 2007).

[7] The idea of generative entrenchment is first developed by Rupert Riedl (1978), and later independently by Arthur in 1984 (1997, 2005) and by me (Wimsatt 1981b, 1986), though the joint work with Schank represents the only attempt to test it through simulation (in models of gene control networks).

[8] Most mutations with large effects used for easy visibility in classical genetics were similarly very deleterious, something that leads to early assumptions that Mendelism and Darwinism were necessarily opposed. (See Provine 1971).

The second part of this argument is that a relatively high proportion of small mutations should be adaptive. This is due to R. A. Fisher (1930) and nicely exploited and extended by Wallace Arthur (1984). Adopting Sewall Wright's (1932) picture of an adaptive topography, Fisher asks us to assume that the surface is continuous and not too bumpy and that we are near to but not at an adaptive peak. Mutations would move us in this adaptive topography. If the surface is continuous and has a nonzero slope, then for very small perturbations from a point, in the limit as their size approaches 0, half of the changes should be adaptive. (Assuming at the point a tangent circle of small radius placed on the surface, half of this will fall "uphill" from the constant fitness isocline through the point and half below.) If one is already near an adaptive peak, as the radius of the "mutation circle" increases, more and more of the area within the circle will be lower than the starting point, so a larger proportion of mutations would be maladaptive, suggesting the outcome of the first argument in another way. If we add together mutations of all sizes, the cumulative distribution would have a much smaller fraction of adaptive variants than the small ones alone. If we reflect that adaptation takes place in a high-dimensional space, this becomes another variant of the preceding argument against the adoption of larger mutations, but what is surprising is that it has a limit for small mutations that is reasonably high. Both of these are plausible arguments, and likely quite robust, though neither of these is rooted in the underlying molecular processes. This disadvantage is not shared by the convergent arguments and data offered by Wagner (2005) that I consider below.

I argued above that larger randomly distributed changes would have a much higher chance of being deleterious than smaller ones. To assume randomly distributed effects is problematic however. Current discussions increasingly note that variations are often multiply correlated in ways that are biased, developmentally canalized, and even if not such as to guarantee success, at least probabilistically biased away from being significantly maladaptive (e.g., Arthur 2004; Kirschner and Gerhart 2005; Wagner 2005). Thus, all of the phenomena of allometric growth involve scaling relations so that as size grows, leg cross sections and intestinal surface area grow in the proportions necessary (as the 3/2 power of length) to preserve adaptive surface-volume ratios for the increased volume that must be gravitationally and metabolically supported. Mutations tend to change these all in a coordinated fashion. Preservation of allometric relations would tend to preserve functional relationships across size changes. This is a lovely case because there must be adaptations to preserve allometry for growth occurring during development, but the adaptations necessary to do this also make the system "preadapted" (or exaptive) to tolerate evolutionary changes in adult size. Darwin noted more generally a "law of correlation of growth"—of which this would have been a special case, in a formulation simultaneously embracing (from our perspective) both genetic pleiotropy (multiple effects of change in single genes) and the adaptive correlations necessary to maintain adequate function. These would become unbalanced should one factor grow substantially without coordinated changes elsewhere. A single mutation that would give a zebra a neck like a giraffe would be a disaster: the forelegs (bones and muscles) would not be able to support the weight, the nervous system would need recalibration for a different gait and balance, the blood pressure would not be sufficient to give the brain a sufficient supply, and so on virtually *ad infinitum*.

Another way to get away with larger changes is if their effects can be bounded in a modular fashion. If interactions with the larger system are at least initially limited, then the number of things that must be done "right" is smaller, and the chance of success greater, though presumably not large. At the character level, this is what is behind Lewontin's (1978) suggestion of the importance of "quasi-independence"—that one should be able to make small changes in one character without affecting others. But this argument should work not only for characters but also at the level of system organization: increased modularity should increase the possible rate of evolution. This kind of situation is found in aggregation, both in going up a level (Simon's (1962) evolution by means of aggregation of stable sub-assemblies), but also in parasitism and symbiosis as is indicated in the fusion of a simpler cell with a mitochondrial ancestor to produce a eukaryotic cell (Sterelny 2004). These will generally become more richly interactive and coevolved later, but the initial modularity makes the aggregation possible (Wimsatt 1974/2007, Ch. 9). Griesemer (2007, in process) has pointed out that such aggregative interactions would generally require supporting structures or dynamics external to both of them. In any case, here one would expect each module to retain its internal functional organization in most respects, though retuning its responses cooperatively, so as to be no longer incrementing fitness only for itself, but for the larger unit of which it is a part. (Contrary to most suppositions, most behaviors benefiting one would also benefit the other, but the problem remains how to avoid or prevent "free riding" for the others.) This is nonetheless a difficult transition, with different theoretical opinions (Maynard Smith and Szathmary 1995), and only now recently investigated in a series of related organisms capable of yielding the detailed steps to cooperativity (Herron and Michod 2008). Finally, parasitism is frequent in evolution and, indeed, has likely been an important driver of increases in adaptive variety in the host (Ray 1991) (because heterogeneity makes a population or community less invasible to a specialized predator, Wills 1996) and has engendered complex specialized and generalized adaptations (of native and acquired immunity) to resist foreign invasion.

So evolution, with few exceptions generally favors smaller changes without ruling out (successively rarer) adaptive changes of successively larger size. We have also considered changes in functional organization with allometric growth (which tends to minimize such changes), and symbiosis or parasitism. But we still need to consider further two important questions. The first is, how does this discussion of magnitude of changes map onto what sorts of changes might be expected in *functional* elements or broader *functional* organization? The magnitude of change refers to differences in fitness that are overall measures of the efficacy of functional organization but totally "black-boxes" how this is accomplished. And secondly, how might functional organization constrain changes—especially deep or far-reaching changes—if they are to be adaptive or even survivable?

4 Generative Entrenchment and the Stability of Deep Functions

It is likely that at least the deeper features of any existing functional architecture will have greater antiquity, generality, adaptive, and even conceptual necessity because their causal depth and entrenchment will make them stabler than more surface features. This likelihood grows with the complexity of that relationship for three reasons:

First, a complex structure will have dependency relations that anchor in elements of the structure because other things depend upon them—they are generatively entrenched, maintained by stabilizing selection, and made evolutionarily conservative by virtue of the substantial adaptive value of the things they help to generate, thus a *generative* entrenchment.
Second, "nature does nothing in vain": if something is complex, it is likely very important. If so, it is thereby also likely that there is significant functional redundancy for accomplishing what it does. Its functional performance becomes *robust* (Wagner 2005; Wimsatt 1981a, 2007a).
Third, these two factors interact: things that are evolutionarily stable because they are important are readily presupposed by other new additions that come later, through the constancy of these stabler elements, thus increasing their generative entrenchment. And things that are entrenched are things for which it is advantageous if they become increasingly robustly generated. Thus, entrenchment and robustness would tend to feed upon each other (Wimsatt 2003).

These are three natural factors in the evolution of complex organization, but a *fourth* conceptual factor is also important. In the limit, when these aspects of function become absolutely central and crucial, they come to be taken as constitutive of the kind of object that it is. (One might say that they become *conceptually* entrenched and come to have even a definitional role in characterizing that kind of object.) This seems like a point just about taxonomy, but this kind of relationship would have made Ernest Nagel's original analysis for functional elements as giving them a *necessary* role in *proper* functioning plausible (Nagel 1961) though Nagel's account is mute as to why this should be so. This analysis provides an explanation. In biology, this is the stuff of fundamental architectures of higher taxonomic categories or *Bauplans.*

Finally, even though these four considerations arise in the context of a selectionist analysis of function, they also affect Cummins' *role* function, for there too (were advocates of this style analysis to talk about it—they commonly don't), *difficulty* of change induces stability over time.

These factors give us a general bias toward stasis for major features, but don't talk about how mechanisms of change will play out at the more specific and concrete *functional* level.[9] Since the paper of Gould and Vrba (1982), there has been

[9] Gould's (1977) interest in heterochrony fits as an attempt, through analyzing changes in relative timing of different developmental systems, to give a general characterization of many such changes.

increased awareness of how commonly things selected for one function are co-opted for another. They may retain the original function or with sufficient redundancy, may even lose it (thought not usually immediately) as it is elaborated in new directions. Swim bladders become lungs. Scales become hair. Bones in an articulated fish jaw become stirrup bones in an ear. Forelimbs become articulated hands with opposable thumbs. Olfactory cortex in our ratlike placental mammal ancestors effloresces into the far larger and more elaborate higher cortex in humans. Ballistic calculators become life-insurance data processors; become scientific minicomputers; become PCs; become home appliances for playing DVDs, cellular telephones for surfing the Internet, and music players; and multiply to perform and integrate hundreds of distributed functions and diagnostic procedures in automobiles. All of these suggest co-option but in ways that are possible at multiple functional levels.

With close inspection one can usually track semicontinuous transitions from one function and functional architecture to another at most junctures, but why are these architectures stable at all? Why isn't the new change in function sufficient to cause (or to require) a total morphing into a new functional architecture? Must small mechanical changes map to small functional changes? Or are there strong tendencies in that direction? We are far short of an exhaustive answer to this question, but there are several considerations bearing on it:

1. "Small change" means "small change in fitness" which is usually most easily accomplished via a small change in function. If a small material change produced a large change in function, it would likely lead to a large change in fitness and subject to most of the same constraints, rendering larger adaptive changes improbable.
2. "Correlations of growth" build into the developmental architecture of the phenotype produce more invariance of functional organization than would be expected at random.
3. Aggregation of modules should commonly preserve most aspects of functional organization if that function is locally characterized. Since a module is now included in a larger system, higher-level or more distant functional consequences might be expected to be more changeable than local ones concerning the interior operation of the module but often to be a combinatorial function of these modules.
4. To be co-opted for another function economically, the functional element or system usually already has the structure and organization to perform the new function and is probably already doing it at some level of adequacy—one sufficient to make co-opting it useful. This already suggests that retuning involving relatively modest morphing of that structure by changing relative proportions or modes of interaction by modest amounts could serve to improve function for the new task for which it would not have been optimized or satisficed already. So there are very likely to be nearby incremental improvements that do not change functional architecture or change it only slightly.
5. So why not make a big change? Why be conservative? The reason for this has been covered but deserves reemphasis. The elaboration of existing function must take place while preserving adequate performance of the prior function. Unless

the system is already redundant (as with tandem-duplicated genes or larger genetic units in systems that permit this), the existing unit must continue to do so.

6. The preceding fact directs attention to organizations that duplicate functions or have redundant capability for realizing them in some degree in a possibly distributed fashion by entities that were doing something slightly different. This is one clear way in which change can happen though usually it does so slowly.
7. It may be that systems whose functional architecture changes too readily when placed in new contexts or when they are perturbed slightly are not sufficiently robust in their behavior to be evolutionarily useful or stable. Fragile designs don't make good evolutionary sense, since presumably a larger fraction of mutations affecting them would be deleterious. This raises the question whether there is in effect selection for robustness of, and thus relative stability of, functional architectures.[10]

5 Multiple Realization, Stability, Robustness, and Evolvability

Stability of functional organization has been taken up in the philosophical literature largely under the headings of multiple realizability or functional equivalence. The largest defect of these discussions for our purposes is that they have not at all addressed issues of change of function or functional organization, or of the robustness of the posited multiple realizability. Abstract discussions of the existence of multiple realizability by philosophers have tended to suppose that it is widely found, especially for mental functions, and in discussions totally divorced from reality, have treated the problem as analogous to the observation that there are a denumerable infinity of (mathematical) functions through any finite set of data points. It has been used as an argument against reductionism. I find such abstract discussions unproductive, and it is also unsound as an argument against reductive explanations, since (approximate) multiple realizability is a natural consequence of the existence of multiple levels of organization (Wimsatt 1994, 2006; Batterman 2000).

Moving away from the abstract discussions, Bechtel and Mundale (1999) have argued that the clean multiple realizability assumed in philosophical discussions is not found in the neurophysiological phenomena—that upon closer look, it is never exact. This makes it clear that we really need to be looking at how often changes result in systems that are nearly the same, or nearly the same in certain respects, or nearly the same in certain environments. This topic is also discussed in evolutionary genetics under another name. Neutral mutations, mutations of equivalent contributions to fitness or equivalent function are now commonly assumed features in population

[10] Clearly this presupposes a kind of dedicated architecture; not one designed to be trainable and adaptable to a variety of diverse circumstances that call for qualitatively different responses like found for cortical functions. Even here of course certain architectural details must be preserved for proper functioning.

genetic modeling. Though neutrality is now more commonly assumed as a kind of "null hypothesis" used to better detect deviations from it (Kreitman 2004), phenomena emerging from "near neutrality" are increasingly matters of active research interest.[11]

Why have neutrality or near neutrality in evolutionary systems? Consider the following paradox: in sexual species, half of the genome is contributed by each parent. Genetic variation has been found at a significant fraction of loci, and this variation is scrambled in successive generations through recombination. As a result, essentially all individuals are new genetic combinations that have never been seen before. This would be no problem if gene effects were commonly additive. But genes show significant epistasis, or gene interaction, in which changes at one locus affect gene expression at other loci in nonlinear complex combinatorial ways. Losses or additions or rearrangements of relatively small amounts of genetic material, even small parts of a chromosome, can produce significant genetic anomalies and strongly deleterious consequences for fitness. This suggests that there ought to be unpredictable "nonadditive" or "emergent" consequences of new genetic combinations. This makes it first of all astounding that most new zygotes aren't immediately inviable. Remarkably, the estimate of viable zygotes for young reproductive human mothers is of the order of 50% and comparable for other vertebrates. That it is this high already suggests organization to tolerate a kind of genetic scrambling under sexual reproduction and recombination. (Imagine even quite highly constrained random recombinations of parts of computer programs, and you begin to appreciate how remarkable this is). People working in Artificial Life have tried to duplicate it, rarely successfully, for reasons partially explained by Ray (1991). But it is not just viability that is amazing. Strikingly, despite all of the nonlinear gene interactions and new combinations, most offspring phenotypes not only are clearly of the same kind as the parents but reflect their many detailed morphological, behavioral, and biochemical characters. We say, easily, that he has his mother's mouth, his father's eye color, his grandmother's asthma, and his grandfather's sense of humor. Some of these may well be partially culturally mediated, but if there were not significant heritability both of fitness and of individual characters from parents to offspring, evolution would not be possible, and we would not be here. This plausibly extends to dozens or hundreds, even thousands, of traits at any given time in any lineage. But how is this heritability possible in the face of so much variation and epistasis?

[11] Wimsatt and Schank (2004) show that the presence of "nearly neutral" mutations in complex genomes can have quite unintuitive effects in systems with truncation selection. This presumably applies to all complex systems with significant genetic load. Loss of alleles causing weakly deleterious fitness depression can amplify the relative fitness contribution of alleles making larger contributions sufficiently to make them conditionally ineliminable, in effect amplifying resistance to loss through stabilizing selection acting on generatively entrenched traits. Favorable mutations can reduce the relative fitness contributions of other alleles, giving "breathing space" to make them more readily lost or changed, in effect releasing a burst of new variation.

The answer lies in the considerable robustness of most phenotypic traits, something significantly driven by the constant reassortment of sexual reproduction. Wagner (2005) argues that we need substantial robustness to be able to survive environmental diversity, and that in consequence, we also have significant robustness to genetic diversity.[12] Particularly interesting is Wagner's development of the idea of a "neutral mutation space" (Chapter 13), in which trait changes at that level of organization produce phenotypes of equivalent fitness and usually equivalent function. I believe that *this idea is the most productive formulation of ideas of multiple realizability for evolutionary biology and most plausibly also for philosophy of psychology* (Fig. 1).

This idea is first employed by Huynen et al. (1996) in their analysis of the effects of sequence mutations on the folding of RNA molecules. They found (to a first approximation) substantial multiple realizability for a small number of distinct structural forms that could be presumed to affect function—indeed, that a very small number of the more frequent forms covered the major fraction of the space of alternatives. Furthermore, most of these forms had "nearest neighbors" of the same form (often only a single base mutation away) in the sequence space and that they were connected so that one could "percolate" through the state space preserving approximately the same form for substantial distances. Finally, they found that most of the most frequent types had each other as relatively close neighbors in most parts of the space. The net effect is that the sequence could "drift" in the neutral space with relatively little effect until an environmental change (or genetic change elsewhere in the genome) changed expression of an existing variant or made another variant adaptive, with the result that subsequent selection had quite different consequences from what they would have where it started. Populations would diffuse in such a "neutral percolation space" while still remaining phenotypically similar until a new environmental stress would reveal differences in gene expression producing new selectable variation.[13]

With this, Wagner argues that organisms will evolve toward increasing robustness or stability (or canalization) in the expression of phenotypic traits resulting in an apparent paradox: if traits are stabilized in this way, shouldn't robustness decrease the potential for evolution or reduce "evolvability"? Wagner argues that this consequence is avoided because increased robustness is secured via the evolution of, or migration into, a larger region in the "neutral mutation

[12] He suggests that environmental robustness must be the main driver of increases in robustness because he thinks that environmental fluctuations are far more common than mutations. But significant epistatic effects should make sexual recombination, which is orders of magnitude more frequent than mutation, a potent driver for robustness over most changes in genetic environment and not at all obviously less important than environmental fluctuations.

[13] In a striking new study (Isalan et al. 2008) new (promoter) links were added at random to a bacterial genome of *E. coli*. In 586 random trials, 95% of the variations made only relatively small fitness variations, suggesting remarkable robustness and evolvability of major expression patterns.

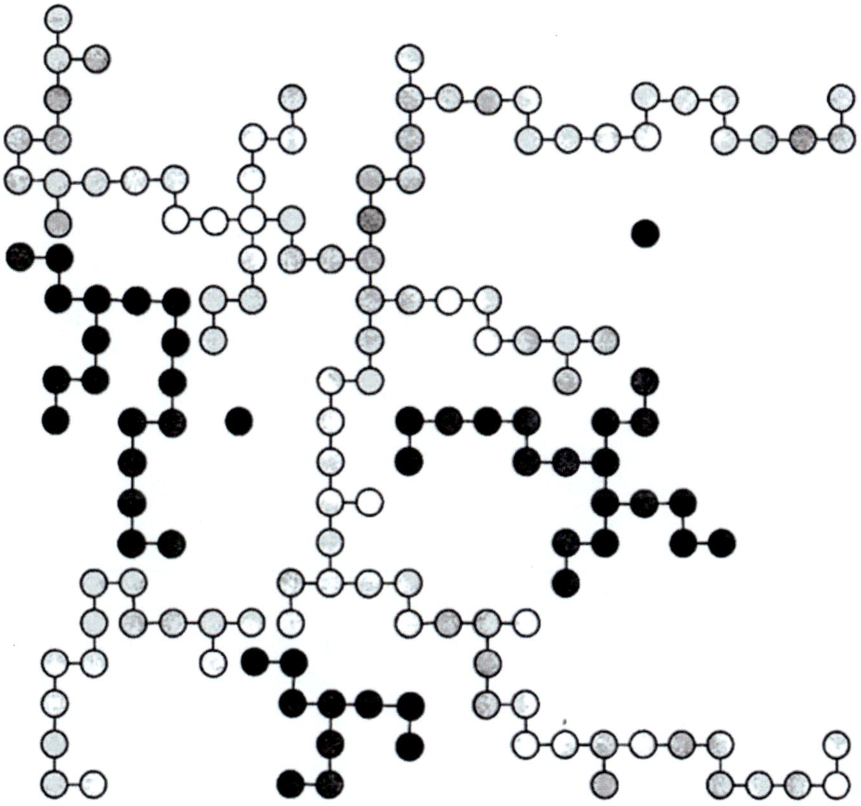

Fig. 1 Schematic illustration of three different classes of RNA or protein structures. The *rectangular area* symbolizes sequence space. *Circles* correspond to individual sequences in this space, and *circles* with the same *shading* correspond to sequences folding into the same (secondary or tertiary) structure. The network of *circles shaded in light gray* corresponds to a highly frequent structure, a structure realized by many sequences. All sequences folding into this highly frequent structure form a connected, neutral network (491). The three groups of *circles shaded in dark gray* correspond to sequences folding into the same moderately frequent structure. The sequences folding into this structure do not form a connected network, but instead form three disjoint sets of sequences. Finally, the *two black circles* correspond to a rare structure, a structure realized by only two sequences that occur at different points in sequence space. The image is misleading in that the actual sequence space is high-dimensional, not two dimensional as suggested by the *box* (From Andreas Wagner 2005, p. 201)

space" for the robust characters.[14] (The robustness of a state is the probability that it will stay the same under perturbation, so can be characterized as the size of the compact state-space region within which it is invariant. Its evolutionary

[14] Wagner supposes that the space remains unchanged by mutations that he represents as leading to migration through it. Myers (2008) points out not only that one must look for sequence changes that leave the function(s) of the protein unchanged but that the proteins must also remain distinguishable from other proteins with which it could be confused, so that it is a coevolutionary problem and this would tend to change the shape of the neutral mutation surface. Wagner's formulation should be regarded as a reasonable first approximation to visualizing the problem but only that.

robustness is its probability of preserving this under migrations out of the immediate region.) But this means that a larger range of states are accessible,[15] and thus, with changed conditions, a larger diversity of variations may be produced, so there is no longer any necessary tradeoff between robustness and evolvability or variety.

So how does this bear on the preservation of functional organization? As I see it, the features of functional organization change in the following ways:

1. In the early stages, self-organization interacts with contingency to produce a workable structure that is not radically improbable (self-organization, *sensu* Kauffman 1993)[16] which is then modulated and tuned for improved functionality through differential selection.
2. Functional subunits may be combined through different lineages. Thus prokaryotic-like cells combined with the ancestors of mitochondria and other initially parasitic endosymbionts yielding the oxygen-based metabolism and richer cellular structures of eukaryotes. Similarly the four-wheeled wagons were combined with steam (inspired by locomotives) and later with internal combustion power plants to produce automobiles and trucks, which are then coevolved and elaborated for better coarticulation and efficiency.
3. As modulations and tunings are added and existing structures are co-opted and morphed for different functions, the more deeply entrenched features become relatively stable at that functional level, though they may evolve both at lower levels (through functionally equivalent substitutions in synonymous genes for functionally equivalent gene expression structures) and at higher levels (as they are co-opted for different functions in higher-level systems). And this is of course going on at multiple levels simultaneously.
4. Higher-level embeddings may make lower-level things freer to vary, if, for example, the higher level of organization generates functional redundancy or robustness at the lower level (Gilbert et al. 1996).
5. By the same token, lower-level standardization and fixity may allow a combinatorial explosion of higher-level variability as common standards and interchangeable parts made possible the explosion of mechanical devices constructed by engineers using "off the shelf" parts in different combinations. Some parts of the system may become variable or elaborated differently in different lineages just as other parts become more fixed (Davidson and Erwin 2006).

[15] Unless the state space is changed through evolution (which it must be on a longer time scale) these points on the trajectory will all be connected and thus the system will just be at a different place in the neutral mutation space. But then "accessible" must here be interpreted as *locally* accessible. This is thus an expression of the themes I elaborated in Wimsatt (2007).

[16] Kauffman commonly opposes self-organization and selection. This is an error that we discuss in Schank and Wimsatt (1988). Most commonly, both would be expected to be occurring and usually symbiotically.

6. Some transformations involve scaled relations to preserve adaptive relations. Here the fact that metazoan organisms must manage scale changes in development can exapt them to tolerate significant changes in evolution using the same scaling systems over broader or different ranges to produce generally adaptive allometric transformations.

6 Deep Function and the Limitations of a Selectionist Account of Function

The picture of functional organization and its response to selection that emerges after we take generative entrenchment and robustness into account is different and revealing. Small changes—the "individual variations" that Darwin took to drive evolution, especially when they are single alleles that don't affect fitness much —are the stuff of population genetics and microevolution. Single-allelic larger changes are far less frequently adaptive but occasionally significant. The HbS mutation that confers resistance to malaria could be representative of this type. Multiallelic larger changes, according to common wisdom, require additive or at least monotonic effects so that evolution can take a path of multiple successive improvements. This then decomposes to a succession of single-locus events.[17] Multilocus epistatic events are not impossible but require the coordination of several fortuitous events (e.g., the co-occurrence of the right small collection of mutations in a temporarily isolated subpopulation) within a relatively small number of generations so that the first-occurring ones are not lost before the remaining ones occur. If we ask how selection explains function in these contexts, it looks like we must opt for the recent history interpretations of the selectionist account (e.g., Godfrey-Smith 1994) or else suppose at least largely directional selection for a geologically extended period of time—something that becomes less likely the longer the period. So it is in the shorter periods and with changes of smaller effect that the selectionist account of function is at its best. But of course, save for single gene substitutions, we would rarely see the creation of new functional systems in short periods, but instead see mostly fine-tuning and elaboration of existing systems. This leaves only relatively thin "additions" to function as what is incrementally selected for.

[17] This is actually too strong a condition as is demonstrated by multi-locus A-Life simulations on the evolution of functional organization by Lenski et al. (2003). They looked at the trajectories that led to adaptive solutions and found that 43% of the steps in ultimately successful trajectories were actually to states of lower fitness than the preceding. If a state is only slightly less fit, it may be able to survive for significant time before it goes extinct—enough time to mutate to a fitter state than the original. This possibility would be recognized in principle but generally ignored in discussions and models. What is striking about their results is just how frequently this occurs—surely much more frequently than most of us would have supposed, as is reflected by the rarity with which this possibility is even considered in the literature.

As we move to deeper aspects of functional organization, the sort that mark differences in *Bauplan* (e.g., the endoskeleton of all chordates versus the exoskeleton of arthropods), there are no single-allele mutations that can switch from one to the other, and the effects of generative entrenchment tell us that it could not possibly be adaptive to do so even if there were such mutations. (The famous "bithorax" mutation that duplicated the winged body segment in Drosophila and led the way to the discovery of HOX genes was in some ways a throwback to the four-winged ancestors of the more modern dipteran flies (which originated about 240 mya), but it was not a viable mutant in any but a modern genetics lab, where its interest led to its careful husbandry.) This kind of change in functional organization runs so deep that there is no adaptive—even no meaningful—transition from one to the other.

This puts the contrast between role function and selection function in a new light. It is not just that people investigating role functions are not *interested* in a selectionist account of their origin. They could well be at a suitable distance. Rather, the problem is that *there is no differential selectionist account moving from any form that is anywhere close to the form under consideration to the alternative modes of organization considered in a comparative morphological analysis.* Considering, for example, the transition from four-winged flies to the Diptera, or even more extremely, the exoskeleton of arthropods versus the endoskeleton of vertebrates, there is no adaptive variability in the population for those sorts of changes because they are all lethal, and any such variability would have been so far in the evolutionary past as to be between variants none of which would have been recognizable as arthropods or chordates. This has another relevant implication: *the reason any mutations deep enough to affect the architecture of an endoskeleton or exoskeleton are lethal is because they affect so many other elements. This explains the absence of viable mutants through generative entrenchment. But the far-reaching consequences of generative entrenchment generate a kind of incommensurability: any possible transitions would affect so many other systems that it would be impossible to localize the functional effects or distribute praise to that system alone.* (This problem is bad enough for pleiotropic genes. It is enormously magnified for traits so deep that they are significantly canalized and disturbed only through perturbations that wreak havoc elsewhere in the system.) I noted this in 1972 as a problem for picking a reference situation for comparison in evaluating fitness claims and also noted that in effect there was a problem of incomparability (incommensurability) for functional systems that were too different. I think now that the problem must be recognized as deeper and more far-reaching.

So given the absence for such "deep" functions of a differential selectionist transition to the state having the function from a state of another organism lacking it, for the kind of comparisons drawn in functional morphology or macroevolutionary contexts, it is reasonable that "causal role functionalists" should resist a differential selectionist account in unpacking their causal role comparisons. In this situation, a causal role account of how the mechanism works can provide in outline the reasons why it could have been elaborated (along its now separate track) to its current functional state from the last common ancestor with its comparison organism. As such, it keys into a selectionist narrative, but not a differential selectionist one—even though it would have originated through processes of differential selection (and drift, and isolation, and all of the other relevant forces and factors.) But the more different they are, the more difficult it is to conceive of them usefully as true alternatives. To the extent that there are comparisons with alternative designs at this molar level, they can point to different conditions appropriate to evaluating the different designs, but this is in effect to accept that they are being evaluated relative to different niches.[18] Note, however, that as adaptive designs, this account is still referred broadly to selectionist contexts, and one could not thereby plausibly choose to give a causal role description that would be inconsistent with this etiology. But it is no longer a differential selection account that is given. *So the idea that the causal role descriptions can be unconstrained by biological ends seems incorrect. Only by picking them out of context might causal role descriptions seem to have this freedom.* But there is also an intelligible sense to saying that they are not to be referred to a *differential* selection or a population genetic account for their validation or that there is such an account in their history.

Lauder and Amundson (1994) suggest that many functional morphologists conceptualize the function problem without any reference to environment. When this is so, it suggests that the description of function has become abstracted and generalized (by comparison of similarities across different contexts) sufficiently to remove any overt reference to environment, or perhaps only to such a generalized environment that it appears possible to leave it unstated. (How, e.g., for a locomoting vertebrate, do you specify particular environments in which having a hemispherical cartilage-to-cartilage joint is adaptive, since it is required for any possible changes in limb direction or extension?) One of the intriguing parallels here is that the more deeply entrenched a functional trait becomes, the more unconditionally deleterious is its change or loss. Here, too, the relevance of context to function is erased, even though fitness and function are both essentially contextual notions.

[18] This still makes it possible to argue that at least some differences of the form one might consider as the subject of comparative morphology could be subject to a differential selectionist interpretation, but this would be situations in which there was competition between different species. Thus, the displacement of many marsupial organisms following the introduction of placental mammals to the Australian ecosystems by Europeans could support a selectionist account. But the reasons for many other interspecies replacements (e.g., the mass decimation of ground-nesting native birds on many Pacific islands) would not be referred to superior performance in broadly similar but detailed different functional systems.

7 Two Modes of Descriptive Abstraction for Function

This suggests another broader principle that also accompanies the knowledge that has come with the discovery of wide reuse of genes in different contexts. *As we attempt to characterize their functions, I suggest that the desire to generalize across contexts leads us to specify what they do in one of two ways:*

1. *We may specify them more locally, removing broader contexts and consequences from their functional descriptions and attributions.* This modularization of description of function can also act as a better tool for understanding how they can act as modular elements in a broader range of contexts and could, for example, even facilitate prediction and design in genetic engineering. Such a decontextualization of function could make the functions attended to appear to be more evolutionarily stable and more general while becoming more modular. In these situations, we should expect the same part to be performing the same function again and again in increasing numbers of local contexts as it is modularly reused, but this is in part because we are decontextualizing its activity. This practice of giving more local characterizations of function is done so at the cost of decoupling its functional description from its specific role in the larger functional organization, but might suggest, in its decontextualization, the Cummins' role account
2. Alternatively, one could abstract the function, removing details of its operation, producing successively higher-level accounts. Such an abstraction of function ought also to increase the apparent stability of function in evolutionary time and ought to increase its apparent generality. So thus entrenchment, robustness, modularity, generality, and descriptive level are all implicated in the real and apparent stability, through time, of functional roles. It should be harder, however, to maintain this stability as we look at successively larger chunks of a functionally organized system.

8 Conclusion

We see that functional organization has a number of factors leading to its relative stability or constraining its evolution. These factors, particularly generative entrenchment and robustness, place significant constraints on its form. These also interact with our practice in how we define function so that the net effect is that there are both biological (or technological, for intentionally constructed material artifacts and plans) and cognitive or conceptual factors dealing with how we generalize or abstract functions that also affect the perceived stability of functional organization. Finally, we see that the form of evolution of complex structures and particularly the conservatism of deeper structures give reason for distinguishing uses of function that arise in the context of evolutionary explanations involving differential selection from those used in comparative or functional morphology, though both are equally causal, and both are to be referred ultimately to the elaboration of organization through selective forces. I think it unlikely that we yet have a complete understanding of the factors tending to generate relative stability (or transformation) of functional forms, and some of the causes will surely remain local and contingent and contextual. But I am convinced that we have identified several important sources of this stability, both in nature and in our practices in theorizing about and describing it.

Acknowledgement I wish to think Philippe Huneman and Alan Love, whose suggestions have improved this chapter. I encourage Alan to elaborate the additional notion(s) of function that he thinks have not yet been captured by this analysis.

References

Alvarez, L.W., W. Alvarez, F. Asaro, and A.V. Michel. 1980. Extraterrestrial cause for the cretaceous-tertiary extinction. *Science* 208: 1095–1108.

Arthur, W. 1984. *Mechanisms of morphological evolution: A combined genetic, developmental and ecological approach*. New York: Wiley.

Arthur, W. 1997. *The origin of animal body plans: A study in evolutionary developmental biology*. New York: Cambridge University Press.

Arthur, W. 2004. *Biased embryos and evolution*. New York: Cambridge University Press.

Barker, G. 2008. Biological levers and extended adaptationism. *Biology and Philosophy* 23: 1–25.

Basalla, G. 1987. *The evolution of technology*. Cambridge: Cambridge University Press.

Batterman, R.W. 2000. Multiple realizability and universality. *The British Journal for the Philosophy of Science* 51: 115–145.

Bechtel, W., and J. Mundale. 1999. Multiple realizability revisited: Linking cognitive and neural states. *Philosophy of Science* 66(2): 175–207.

Bigelow, R., and R. Pargetter. 1987. Functions. *Journal of Philosophy* 74(4): 184–196.

Campbell, D.T. 1974. Evolutionary epistemology. In *The philosophy of Karl Popper*, vol. 2, ed. P. Schilpp, 412–463. LaSalle: Open Court.

Davidson, E., and D. Erwin. 2006, 10 February. Gene regulatory networks and the evolution of animal body plans. *Science* 311: 796–800.

Dawkins, R. 1976. *The selfish gene*. Oxford: Oxford University Press.

Eldridge, N., and S. Gould. 1972. Punctuated equilibrium: An alternative to phyletic gradualism. In *Models in paleobiology*, ed. T.J.M. Schopf. San Francisco: Freeman, Cooper.

Fisher, R. 1930. *The genetical theory of natural selection*. Oxford: Clarendon Press.

Gilbert, S., J. Opitz, and R. Raff. 1996. Resynthesizing evolutionary and developmental biology. *Developmental Biology* 173: 357–372.

Godfrey-Smith, P. 1994. A modern-history theory of functions. *Nous* 28: 344–362.

Gould, S.J. 1977. *Ontogeny and phylogeny*. Cambridge: Harvard University Press.

Gould, S.J., and E. Vrba. 1982. Exaptation—A missing term in the science of form. *Paleobiology* 8: 4–15.

Grant, P., and R. Grant. 2008. *How and why species multiply: The radiation of Darwin's finches*. Princeton: Princeton University Press.

Griesemer, J.R. (in process) What Simon should have said, lecture, ISHPSSB 2007 (Exeter, July 2007).

Herron, M.D., and R.E. Michod. 2008. Evolution of complexity in the volvocine algae: transitions in individuality through Darwin's eye. *Evolution* 62: 436–451.

Huynen, M., P.F. Stadler, and W. Fontana. 1996. Smoothness within ruggedness: The role of neutrality in adaptation. *Proceedings of National Academy of Science USA* 93: 397–401

Isalan, M., C. Lemerle, K. Michalodimitrakis, C. Horn, P. Beltrao, E. Raineri, M. Garriga-Canut, and L. Serrano. 2008, April 17. Evolvability and hierarchy in rewired bacterial gene networks. *Nature* 452: 840–845.

Kauffman, S.A. 1993. *The origins of order*. Oxford: Oxford University Press.
Kirschner, M., and J. Gerhardt. 2005. *The plausibility of life: Resolving Darwin's dilemma*. New York: John Norton.
Kreitman, M. 2004. The neutral theory is dead: Long live the neutral theory. *BioEssays* 18(8): 678–683.
Lauder, G., and R. Amundson. 1994. Function without purpose: The uses of causal role function in evolutionary biology. *Biology and Philosophy* 9: 443–469.
Lenski, R.E., C. Ofria, R.T. Pennock, and C. Adami. 2003. The evolutionary origin of complex features. *Nature* 423: 139–144.
Lewontin, R.C. 1978. Adaptation. *Scientific American* 239(3): 212–228.
Lewontin, R., and L.C. Dunn. 1960. The evolutionary dynamics of a polymorphism in the house mouse. *Genetics* 45, 705–722.
Lloyd, E.A. 1988. *The structure and confirmation of evolutionary theory*. New York: Greenwood Press.
Maynard Smith, J., and E. Szathmary. 1995. *The major transitions in evolution*. Oxford: Freeman Spektrum.
Millikan, R.G. 1984. *Language, thought, and other biological categories*. Cambridge: MIT Press.
Mills, S., and J. Beatty. 1979. The propensity interpretation of fitness. *Philosophy of Science* 46(2): 263–286.
Myers, C.A. 2008. Satisfiability, sequence niches and molecular codes in cellular signaling. *IET Systems Biology* 2(5): 304–312.
Nagel, E. 1961. *The structure of science*. New York: Harcourt Brace and Jovanovich.
Okasha, S. 2006. *Evolution and the levels of selection*. Oxford: Oxford University Press.
Price, T. 2008. *Speciation in birds*. Greenwood Village: Roberts and Co.
Provine, W.B. 1971. *The origins of theoretical population genetics*. Chicago: University of Chicago Press.
Raup, D. 1993. *Extinction: Bad genes or bad luck*. New York: W. W. Norton.
Ray, T.S. 1991. Is it alive, or is it GA? In *Proceedings of the 1991 International Conference on Genetic Algorithms*, ed. R.K. Belew and L.B. Booker, 527–534. San Mateo: Morgan Kaufmann.
Riedl, R. 1978. *Order in living organisms: A systems analysis of evolution*. New York: Wiley.
Schank, J.C., and W.C. Wimsatt. 1988. Generative entrenchment and evolution. In *PSA-1986*, vol. 2, ed. A. Fine and P.K. Machamer, 33–60. East Lansing: The Philosophy of Science Association.
Simon, H.A. 1962/1996. The architecture of complexity. Reprinted in his *The Sciences of the Artificial*, 3rd ed. Cambridge: MIT Press.
Sterelny, K. 2004. Symbiosis, evolvability and modularity. In *Modularity in evolution and development*, ed. G. Schlosser and G. Wagner, 490–516. Chicago: University of Chicago Press.
Wagner, A. 2005. *Robustness and evolvability in living systems*. Princeton: Princeton University Press.
Wills, C. 1996. *Yellow fever, black goddess: The coevolution of plagues and peoples*. New York: Harper Collins.
Wimsatt, W.C. 1971. Some problems with the concept of feedback. In *PSA-1970*, Boston studies in the philosophy of science, vol. 8, ed. R.C. Buck and R.S. Cohen, 241–256. Dordrecht: Reidel.
Wimsatt, W.C. 1972. Teleology and the logical structure of function statements. *Studies in History and Philosophy of Science* 3: 1–80.
Wimsatt, W.C. 1974. Complexity and organization. In *PSA-1972*, Boston studies in the philosophy of science, vol. 20, eds. K.F. Schaffner and R.S. Cohen, 67–86. Dordrecht: Reidel, Reprinted as ch. 9 in Wimsatt, 2007b.
Wimsatt, W.C. 1980. Reductionistic research strategies and their biases in the units of selection controversy. In *Scientific discovery*, Case studies, vol. 2, ed. T. Nickles, 213–259. Dordrecht: Reidel.
Wimsatt, W.C. 1981a. Robustness. Reliability and overdetermination. In *Scientific inquiry and the social sciences*, ed. M. Brewer and B. Collins, 124–163. San Francisco: Jossey-Bass.
Wimsatt, W.C. 1981b. Units of selection and the structure of the multi-level genome. In *PSA-1980*, Lansing, vol. 2, ed. P.D. Asquith and R.N. Giere, 122–183. Michigan: The Philosophy of Science Association.

Wimsatt, W.C. 1986. Developmental constraints, generative entrenchment, and the innate-acquired distinction. In *Integrating scientific disciplines*, ed. P.W. Bechtel, 185–208. Dordrecht: Martinus-Nijhoff.
Wimsatt, W.C. 1994. The ontology of complex systems: Levels, perspectives and causal thickets. *Canadian Journal of Philosophy supplementary* 20: 207–274.
Wimsatt, W.C. 2001. Generative entrenchment and the developmental systems approach to evolutionary processes. In *Cycles of contingency: Developmental systems and evolution*, ed. S. Oyama, R. Gray, and P. Griffiths, 219–237. Cambridge: MIT Press.
Wimsatt, W.C. 2002. Functional organization, functional inference, and functional analogy. In *Functions: New essays in the philosophy of psychology and biology*, ed. Robert Cummins, Andre Ariew, and Mark Perlman, 174–221. Oxford: Oxford University Press.
Wimsatt, W.C. 2003. Evolution, entrenchment, and innateness. In *Reductionism and the growth of knowledge*, ed. Terrance Brown and Leslie Smith, 53–81. Mahwah: Lawrence Erlbaum and Associates.
Wimsatt, W.C. 2006. Reductionism and its heuristics: Making methodological reductionism honest. *Synthese* 151: 445–475.
Wimsatt, W.C. 2007a. On building reliable pictures with unreliable data: An evolutionary and developmental coda for the new systems biology. In *Systems biology: Philosophical foundations*, ed. F.C. Boogerd, F.J. Bruggeman, J.-H.S. Hofmeyer, and H.V. Westerhoff, 103–120. Amsterdam: Reed-Elsevier.
Wimsatt, W.C. 2007b. *Re-engineering philosophy for limited beings: Piecewise approximations to reality*. Cambridge, MA: Harvard University Press.
Wimsatt, W.C., and J.C. Schank. 1988. Two constraints on the evolution of complex adaptations and the means for their avoidance. In *Evolutionary progress*, ed. M. Nitecki, 231–273. Chicago: The University of Chicago Press.
Wimsatt, W.C., and J.C. Schank. 2004. Generative entrenchment, modularity and evolvability: When genic selection meets the whole organism. In *Modularity in evolution and development*, ed. G. Schlosser and G. Wagner, 359–394. Chicago: University of Chicago Press.
Wimsatt, W.C., and J.R. Griesemer. 2007. Reproducing entrenchments to scaffold culture: The central role of development in cultural evolution. In *Integrating evolution and development: From theory to practice*, eds. R. Sansome and R. Brandon, 228–323. Cambridge, MA: MIT Press.
Wright, S. 1932. The roles of mutation, inbreeding, cross-breeding and selection in evolution. *Proceedings of the Sixth Annual Congress on Genetics* 1: 356–366.
Wright, L. 1973. Functions. *Philosophical Review* 82(April): 139–168.

Mechanism, Emergence, and Miscibility: The Autonomy of Evo-Devo

Denis M. Walsh

Abstract Evolutionary developmental biology shows us that the capacities of organisms play an indispensable role in the explanation of adaptive evolution. In particular, the goal-directed properties of organisms figure in a class of emergent teleological explanations. The role of emergent teleology has heretofore gone unnoticed largely because of modern biology's methodological commitment to mechanism. I outline and defend an alternative to mechanism: explanatory emergence. According to explanatory emergence, every phenomenon has a complete mechanistic explanation, yet some phenomena also have emergent teleological explanations. Mechanistic and emergent teleological explanations of the same phenomena are complete, complementary and autonomous. I call this relation 'miscibility'. I argue that the miscibility of explanations illuminates the distinctive character of recent evolutionary developmental biology (evo-devo). Evo-devo offers a class of emergent explanations that advert to the unique capacities of organisms.

Organisms are singular features of the natural world; they are self-organising, self-building, complex adaptive systems. They synthesise the very materials out of which they are constructed. They possess unparalleled capacities for adaptive accommodation to the vagaries of their internal and external conditions of existence. It is natural to suppose that the remarkable capacities of organisms should find some important place in the explanation of biological phenomena. And yet, the category of the organism plays very little role in modern evolutionary biology. Throughout most of the last century and a half, an emphasis on the distinctiveness of organisms has been widely thought to be an impediment to the progress of biology (Allen 2005). A biology set apart by the uniqueness of organisms cannot avail

D.M. Walsh (✉)
University of Toronto, Toronto, ON, Canada

P. Huneman (ed.), *Functions: selection and mechanisms*, Synthese Library, Studies in Epistemology, Logic, Methodology, and Philosophy of Science 363, DOI 10.1007/978-94-007-5304-4_3, itself of the methods and precepts of the physical sciences, nor can it enjoy their cachet. Those methods and precepts include, significantly, a commitment to mechanism. Mechanism is (in part) the view that to explain the properties of a complex system, one appeals to the causal capacities and relations of its parts. Thanks to the pervasive influence of mechanism, biology now needs no special pleading on behalf of the fact that organisms are organisms. Their distinctive capacities are wholly accounted for by the activities of their sub-organismal parts. In this respect, they are no different in kind from any other complex mechanical contrivance. There can be no gainsaying the empirical successes of mechanist-inspired biology, particularly in the twentieth century. But it must be acknowledged that, for better or worse, the resolute pursuit of this program has led to the marginalisation of the organism.

There has recently been a considerable amount of interest in reviving the organism (e.g. Gilbert and Sarkar 2000). Indeed one of the explicit objectives of many practitioners of evolutionary developmental biology (Evo-Devo) is to fashion for organisms an irreducible explanatory role in evolution (Callebaut et al. 2007). Organisms, according to many evo-devotees, are agents of evolutionary change in so far as they actively direct and regulate the origin, development and inheritance of biological form. This program too has been enormously fecund. Few who are acquainted with the startling achievements of evo-devo in the last 20 years can deny that its emphasis on the role of organisms in evolution exposes the sterility of the traditional gene-centred, sub-organismal view of development, inheritance and evolution (Muller 2007). But, this empirical success alone cannot secure a privileged place for organisms in evolutionary biology. The obstacle to organicism is less empirical than methodological. Mechanism stands in the way—at least it appears to. If every biological phenomenon is exhaustively explained by the activities of sub-organismal parts, there is nothing left over for the capacities of whole organisms uniquely to explain.

I believe that, mechanism notwithstanding, the organocentric aspirations of evo-devo can be realised, but doing so requires a significant methodological shift away from unreconstructed mechanism. The position I wish to outline is as follows: Every phenomenon has a complete, mechanistic explanation, but not every genuine explanation is mechanistic. There also exists a class of emergent teleological explanations that appeal to the goals or purposes of a system. The most significant empirical achievement of evo-devo, in my view, has come in demonstrating that the distinctive characteristics of organisms underwrite a battery of these emergent explanations. The pressing methodological challenge facing evo-devo is that of showing how the same biological phenomena may be susceptible of both mechanistic and emergent explanations. I argue that the relation between mechanistic and emergent explanations is not one of mutual exclusion but one I call 'miscibility'. Mechanistic and emergent explanations are miscible in the sense that they offer complete, complementary, autonomous but *different* explanations of the same phenomena. Evo-devo demonstrates that the miscibility of emergent and mechanistic explanations is of vital importance to the explanation of adaptive evolution.

1 Mechanism

Mechanism embodies a simple and compelling idea that to explain a phenomenon, we cite the mechanisms that cause it. It is hardly new; its precursors are found prominently in antiquity. More significantly, it is the methodological standard raised by the scientific revolution. A contemporary variant of mechanism has been vigorously promoted in recent years (Machamer et al. 2000; Bechtel and Richardson 1993; Bechtel and Abramsen 2005; Glennan 1996, 2002; Craver 2007). Not only is this modern variant on mechanism compelling in its own right, it also harbours a crucial insight that I shall exploit in augmenting mechanism with a plausible version of emergentism.

A mechanism is a kind of cause:

> Specifically: Mechanisms are entities and activities organized such that they are productive of regular changes from start or set-up to finish or termination conditions.
>
> Activities are the producers of change....
>
> Entities are the things that engage in activities. (Machamer et al. 2000: 2–3)

A mechanistic explanation works by showing how the characteristic activities of the entities in question produce the effect to be explained:

> [E]xplanation involves revealing the productive relation. It is the unwinding, bonding and breaking that explain protein synthesis; it is the binding, bending, and opening that explain the activity of Na+ channels. It is not the regularities that explain but the activities that sustain the regularities. (Machamer et al. 2000: 22)

We understand a natural phenomenon when we are in possession of a descriptively adequate account given in terms of 'bottom out' activities of the system's parts. A bottom out activity is a behaviour of a 'relatively fundamental' structure that is taken to be unproblematic. A mechanistic account is descriptively adequate in that it is in principle possible to fill in the details of the bottom out activities in such a way as to reveal the phenomenon to be explained as the product of those activities (Machamer et al. 2000).

The stipulation that a mechanistic explanation is *descriptively adequate* often goes unremarked, but it is of particular importance here. An explanation identifies a mechanism, but not just any description of the mechanism is genuinely explanatory. The reason (as I make it out) is that a good explanation answers to both metaphysical and cognitive demands. In a mechanistic explanation, the metaphysical demand is met by identifying the entities and activities that produce the effect to be explained. The cognitive demand is met by describing the system in a way that makes the productive relation intelligible. 'Intelligibility arises not from an explanation's correctness, but rather from an elucidative relation between the explanans and the explanandum' (Machamer et al. 2000: 22). Descriptively adequate mechanistic explanations typically employ 'thick' causal concepts, like *pushing*, *attracting*, *folding*, *pumping* and *compressing*.[1] It is these concepts that 'reveal the productive

[1] I take the expression 'thick' causal concepts from Cartwright (2004).

relations' between mechanism and effect. To give a mechanistic explanation, then, we need two things: (1) a relation between explanans and explanandum and (2) a description of the relation that elucidates its explanatory nature. On account of this second condition, explanation is *description dependent.*[2]

Most mechanists implicitly hold that explanation is more than merely description dependent; it is *description relative*. Explanation is description relative just if for some explanandum, x_e, there is a relation between x_e and an explanans, $x_{c,1}$, and a description, d_1 that illuminates the explanatory nature of the relation *and* a relation between x_e and explanans, $x_{c,2}$, and *another* description d_2 that reveals the explanatory nature of that relation.[3]

This is a commonplace arrangement in the natural sciences (Sober 1999). To choose a simple example, the contraction of striated muscles is explained in two different ways by appeal to two distinct suites of bottom out activities.[4] One explanation focuses on the functional morphology of muscles. A muscle comprises a large number of sarcomeres arranged end to end along the muscle fibre. Each sarcomere consists of alternating rows of long actin and myosin proteins arranged parallel to the long axis of the muscle. Myosin is a thick protein with a golf club-shaped 'head' that can bind with actin. Actin is a thin helical protein. When the muscle is depolarised, the myosin head binds to a site on the actin. The head of the myosin protein bends, which pulls and twists the actin fibre along the myosin fibre. The myosin then releases the actin, resumes its resting position and then repeats the bind-bend-pull-twist-release sequence. In this way, the actin fibre is ratcheted along the myosin fibre, shortening the sarcomere. The effect, summed over the sarcomeres, is a contraction of the muscle.

Binding, bending, pulling and twisting are 'bottom out' activities at this functional-morphological level of explanation. This explanation offered at the level of functional anatomy of actin and myosin proteins is descriptively adequate in the sense that the bending of the myosin head and the twisting and pulling of actin by myosin reveal the way in which the relation between actin and myosin produces the contraction of the sarcomere.

The same process of muscle contraction also has a strictly chemical explanation. The release of ATP in the region of the myosin causes myosin to bind to it. This, in turn, causes a conformational change in the myosin, releasing the myosin head from the actin. The hydrolysis of ATP causes myosin to enter a low-energy state and, hence, to 'unbend'. The release of Ca++ into the muscle exposes the 'next' binding site on the action. Myosin bonds weakly to the actin binding site. The release of inorganic phosphate causes the myosin to bond more strongly to actin and to bend.

[2] The description dependence of explanations is well documented. Davidson (1967), for example, draws our attention to the fact that while causal contexts sustain the intersubstitution of co-referring descriptions *salva veritate*, explanatory contexts do not. The reason is that the explanatory content of an explanation is sensitive to the way the relation between explanans and explanandum is described.

[3] I leave open the possibility that $xc,_1 = xc,_2$. It is, in fact, commonplace that one and the same relation should be susceptible to different explanatory descriptions.

[4] Illingworth (2008) offers a nice overview of muscle function.

At this point, ADP is released from the myosin, enabling it to bind again to free ATP, thus repeating the cycle. The bottom out activities in this explanation—*depolarising, hydrolysing and binding*—are all chemical processes.

The important point is that the same phenomenon—the shortening of the sarcomere—is explained in two different ways, one functional-morphological and the other chemical. Each explanation identifies a mechanistic relation between muscle structure and its ability to contract and elucidates this relation by means of a different set of bottom out activities. Each explanation is in its own right complete, in the sense that each reveals its characteristic activities to be productive of the effect being explained. Any approach that allows that properties at a variety of levels of organisation to enter into explanations of the same phenomenon, under different descriptions, is implicitly committed to description relativity.[5]

Mechanism holds that the properties of a complex entity have a complete mechanistic explanation that adverts to the activities of the system's parts. This in turn has two consequences that together motivate an appealing version of reductionism. The first is that the properties of complex entities have no autonomous causal roles. This is not to say that complex entities do not have causal capacities; they most certainly do. The macrostructures of actin and myosin proteins bind, bend, pull and twist. But every capacity of actin or myosin is vouchsafed by—inherited from—the interactions of their constituent chemicals. The second, corollary, consequence is that the properties of complex entities appear to have no autonomous *explanatory* role. Every effect explained by the capacities of a complex system is also equally explained by the activities of the system's parts. There may be pragmatic reasons for choosing one explanans over the other on an occasion—one may be simpler to understand or remember—but each explains equally well, and in the same way, by citing a descriptively adequate productive cause. We could, for example, replace the functional anatomical description of the action of actin and myosin with a detailed chemical description without any explanatory loss.

Mechanistic reduction of this sort may not be applicable in every discipline or for every (non-fundamental) level of organisation. But, there is no denying that it has been prominent and influential in recent biology. Indeed, twentieth-century biology must be the poster child for reductive mechanism.[6] The modern synthesis theory of evolution is a unification of biological disciplines under the mantle of a single, powerful theory. That theory has come to take a sub-organismal unit of organisation—the replicator—as its canonical entity. Biological phenomena, including development, inheritance and evolutionary change, are comprehensively explained by the activities of replicators. Organisms, now recast as 'vehicles', are more or less incidental to the explanation of evolution:

---

[5] See, for example, Machamer et al. (2000) claim that explanatory properties occur in nested hierarchies. Craver's (2007) mechanistic anti-fundamentalism and Jackson and Pettit's (2004) 'explanatory ecumenism' are also good examples of an implicit commitment to description relativity.

[6] Rosenberg's (2006) spirited and compelling argument is, in many ways, the definitive defence of reductive mechanism in evolutionary biology.

> Evolution is the external and visible manifestation of the survival of alternative replicators … Genes are replicators; organisms … are best not regarded as replicators; they are vehicles in which replicators travel about. Replicator selection is the process by which some replicators survive at the expense of others. (Dawkins 1982: 82)

Replicator mechanics, together with some accumulated chance mutations, wholly explains the process of adaptive evolution:

> …the non-random selection of randomly varying replicating entities by reason of their 'phenotypic' effects … is the only force I know that can, in principle, guide evolution in the direction of adaptive complexity. (Dawkins 1998: 32)

Replicator mechanics, in turn, is explained by appeal to the activities of the molecules from which replicators are constructed (Waters 1996, 2008).

Dissenting voices are being heard (Goodwin 1994; Webster and Goodwin 1996; Callebaut et al. 2007). Evolutionary biologists are increasingly claiming a privileged explanatory status for the capacities of organisms in inheritance (Oyama 1985), development (Muller 2008) and adaptive evolution (West-Eberhard 2003; Muller 2008). The capacities of organisms certainly make important, marvellous and surprising contributions to adaptive evolution, but that alone does not demonstrate the inadequacy of reductive mechanism for evolutionary biology. Many mechanists acknowledge that organisms have wonderful 'emergent' properties that account for important biological and evolutionary phenomena. They further rejoice that mechanism brings these under our ken (e.g. Kauffman 1970; Richardson and Stephan 2007; Bechtel 2007). Our understanding of the mechanics of complex systems has burgeoned in the last 30 years (Kauffman 1993, 1995). This too constitutes a triumph for mechanism.

The success of mechanism raises a problem for the strongly organism-centred program of evo-devo. If every phenomenon has a complete, descriptively adequate (sub-organismal) mechanistic explanation, if there are no explanatory lacunae left for organisms to fill, what can the capacities of organisms uniquely explain? It is in response to this question that we need an account of emergent explanations.

2 Emergence

The contrary of mechanism is usually considered to be emergentism (or emergence). While it is one thing to say what emergentism is not—namely, mechanism—a positive account of the doctrine has proven a little more elusive. Despite the successes of mechanism, emergentism has persisted on the fringes of the philosophy of science (and of certain sciences themselves). I suspect this is partly due to the force of sheer nebulousness: A doctrine that cannot be articulated cannot be refuted. But that is not all that emergentism has going for it. However inchoate it might be, it seems to point toward a genuine insight concerning the proper scientific treatment of complex systems.

2.1 Ontological Versus Explanatory Emergence

Emergentism is usually taken to be a metaphysical doctrine to the effect that the properties or activities of complex systems are importantly different from those of their parts. Some attempts to characterise emergent properties draw a distinction between additive ('resultant') and non-additive ('emergent') effects of the interaction of parts (Alexander 1920). Yet other accounts claim that complex entities have emergent capacities that exert causal influence on their parts (Silbersteen and McGeever 1999). An alternative account of emergent properties has it that complex systems obey laws that their parts do not (Mill 1843). Some accounts characterise emergent properties as those that are insensitive to changes in their microstructural realisers (Rueger 2000). Sometimes, characterisations of emergence have a distinctly epistemic flavour. Complex entities have emergent properties just if they behave in ways that we would not have predicted from a knowledge of their parts taken separately (Broad 1925). Recently, computational criteria of emergence have been suggested, according to which a property *P* is emergent on its microstructural properties, *S*, if *P* can be derived from *S* 'only by simulation' (Bedau 1997) or if its 'generative' explanation cannot be compressed (Bedau 2008; Huneman and Humphreys 2008).

It is certainly true that complex systems have distinctive properties and behave in ways that their parts do not. But that should not cut any ice against mechanism.[7] On the contrary, if mechanism can reveal for us the ways in which the distinctive, unexpected behaviours of complex systems arise out of the interactions of their parts, this should count strongly in its favour. Wimsatt echoes the sentiment:

> Philosophers commonly suppose that emergent properties are irreducible, but some rather nice things fall out of a reductive account of emergence. Claims involving emergent properties in discussions of non-linear dynamics, connectionist modelling, chaos, artificial life, and elsewhere give no support for traditional anti-reductionism or woolly-headed anti-scientism. ... Emergent phenomena like those discussed here are often subject to surprizing and revealing reductionistic explanations. But such explanations do not deny their importance or make them any less emergent – quite the contrary: it explains why and how they are important.... (Wimsatt 2000: 269)

Similar views are expressed by Bechtel and Richardson (1993) and Richardson and Stephan (2007).

As a hedge against this mechanistic reduction, some emergentists claim that the properties of complex entities have substantial causal autonomy (O'Connor 1994, 2000). They claim that complex entities can have causal powers that the concerted actions of their parts cannot. Alas, the prospects for this sort of emergentism are dim. The bugbear for any substantive causal emergence, as Kim has demonstrated, is downward causation. Kim's argument goes as follows.

Emergentists typically hold that the macro-level supervenes upon the micro-level. For any macro-level causal property, C^*, there could be no difference in C^*

[7] In this I agree with Symons (2008) and Bedau (2008).

unless there was also a difference in its microphysical realiser, *C*. Because of this, the causal capacities of *C** are not autonomous from those of *C*. If *C** were to have the capacity to bring about some effect that *C* did not, it would have to be able to cause this new capacity in *C*. That is to say *C** must be able to change *C*, its own microbase realiser. This is reflexive downward causation. Reflexive downward causation is incoherent because in order for *C** to have some capacity that *C* did not have, it would have to be the case that *C* both *lacks* the capacity (otherwise *C** could not cause *C* to have it) and *possesses* it (otherwise, by the supervenience of *C** on *C*, *C** would not have the capacity either). The causal autonomy of emergent properties is inconsistent with their supervenience.[8] Kim's sceptical conclusion is that 'Emergentism cannot live with downward causation, and it cannot live without it. Downward causation is the raison d'etre of emergence, but it may well turn out to be what in the end undermines it' (Kim 2006: 548). Emergent properties may have causal powers, but they have no autonomous causal powers.

Emergentism seems to require that the properties of complex entities have causal autonomy over the capacities of their parts. Mechanism entails that they do not. Thus mechanism and emergentism seem to be incompatible. However, it is not obvious that emergentism really should require causal autonomy of wholes over their parts. After all, if emergentism is an alternative (or complement) to mechanism and mechanism is a doctrine about explanation, then emergentism should be a thesis about explanation too. All emergentism should require, then, is that the properties of complex entities can have *explanatory* autonomy over the properties of their parts. *Explanatory emergentism*, as I shall develop it, is the thesis that the properties of complex entities figure in explanations that cannot be replaced, superseded or augmented by explanations that advert to the activities of the system's parts. An adequate account of explanatory emergence will first need to specify how the capacities of complex entities can figure in autonomous emergent explanations and then demonstrate that explanatory autonomous does not require causal autonomy.

2.2 *Invariance and Explanation*

Nowadays, it is customary to think of explanation in terms of causation: To explain a phenomenon is to cite its cause (Salmon 1984; Lewis 1988; Strevens 2004). The cause of a phenomenon is the set of conditions that makes the difference between its occurrence and its non-occurrence. Difference making is a particular kind of counterfactual relation: x_c is the/a difference maker for x_e just if x_c and x_e instantiate a change-involving invariance relation such that, for a range of circumstances, if one

[8] O'Connor (2000) and Humphreys (1997) attempt to circumvent this problem by arguing that emergent causal properties do not supervene.

were to intervene to change the value of x_c, the value of x_e would also change in a systematic way.[9] Woodward (2003) claims that

> ...the sorts of counterfactuals that matter for purposes of causation and explanation are just such counterfactuals that describe how the value of one variable would change under interventions that change the value of another. Thus, as a rough approximation, a necessary and sufficient condition for *X* to cause *Y* or to figure in a causal explanation of *Y* is that the value of *X* would change under some intervention on *X* in some background circumstances.... (Woodward 2003: 15)

This account of explanation is particularly congenial to mechanists (see Woodward 2002). Being the mechanism of some occurrence is the change-involving invariance relation *par excellence.*

It is important to note here that invariance is not being offered as an analysis of causation. Invariance identifies that relation between a cause and effect that allows us to explain an effect by citing its causes. It is possible, however, that there are other kinds of explanatory relations to which the invariance approach applies equally well. Indeed, we might take this to be the central theme in Aristotle's account of explanation. Aristotle's doctrine of the 'four causes' can be read (anachronistically) as a sophisticated pitch for the multiplicity of explanatory invariance relations. Consider the (admittedly opaque) example from *Physics II*:

> A man is engaged in collecting subscriptions for a feast. He would have gone to such and such a place for the purpose of getting the money, if he had known. He actually went there for another purpose and it was only incidentally that he got his money by going there; and this was not due to the fact that he went there as a rule or necessarily, nor is the end effected (getting the money) a cause present in himself—it belongs to the class of things that are intentional and the result of intelligent deliberation. It is when these conditions are satisfied that the man is said to have gone 'by chance'. If he had gone of deliberate purpose and for the sake of this—if he always or normally went there when he was collecting payments—he would not be said to have gone 'by chance'. (Physics II.5: Aristotle 2007)

The passage is part of Aristotle's account of chance. Its overall objective is to illustrate the poverty of the atomists' approach to explanation (Hankinson 2003). Aristotle charges that the atomists' approach to explanation does not allow them to distinguish an event that occurs 'by chance' from one that occurs as a matter of purpose. This is an explanatorily significant distinction according to Aristotle, and an adequate account of explanation must accommodate it.

Aristotle's atomist opponents believe that the world is made up of a few fundamental kinds of things, atoms, each with a characteristic repertoire of activities. The macroscopic phenomena we observe are produced by aggregates of atoms undergoing their characteristic activities. These phenomena can be explained exclusively by appealing to those activities. This should sound familiar. For all intents and purposes, Aristotle's atomist opponents are relevantly like modern-day mechanists.

---

[9] In the causal modelling literature, 'intervention' has a specific technical meaning. One can intervene on *xc* with respect to *xc* only if there is a direct causal path from *xc* to *xe* (Woodward 2003: 79). I intend to use 'invariance' in a less technical sense. There is a change-involving invariance relation between *xc* and *xe* just if manipulations on *xc* are counterfactually related in the right way to the values of *xe*.

The example suggests that causal/mechanical explanation is insufficiently sensitive to a certain kind of regularity. We can cite all the mechanical causes of our agent's going to the market, his neurophysiological state, the mechanics of his locomotion, etc., right up to the point at which he encounters his subscriber, and there will be nothing in this explanation that tells us whether, under the relevant description ('collecting subscriptions for a feast'), this event is an instance of a regularity or mere chance. The relevant description is one that identifies a purpose for which the event might have occurred. The occurrence happens as a matter of regularity (of the intended sort) only if it occurs *because* it is a means to the attainment of the purpose or goal; otherwise, it occurs as a matter of chance. In the example we are given, the outcome occurs simply as a matter of chance, but the mechanical (efficient cause) explanation cannot tell us that, as the event gets the same causal/mechanical explanation whether it is purposive or not. Aristotle's complaint is that there is a significant class of regular occurrences—those that happen because they contribute to goals or purposes—that atomist/mechanist explanations cannot discern. These regularities are indistinguishable from mere congeries of chance events, unless we cast them under their purposive descriptions. Atomists, according to Aristotle, cannot recognise purposive occurrences as purposive. Consequently, they cannot distinguish them from chance events.

Atomists/mechanists are likely to respond that this is not a chance occurrence. Once we have shown that an event occurred as a matter of causal necessity due to the occurrence of its productive causes, there is nothing more to explain and nothing more to discern. Aristotle's distinction between events that occur because they fulfil goals and those that occur by chance is an empty one.

But this response is surely too doctrinaire. Aristotle is pointing to a structural similarity in the relation that holds between an occurrence and its mechanism, on the one hand, and a goal and its means, on the other. They are both 'change-involving invariance' relations. Goal-directed—purposive—systems have the capacity to bring about states of affairs *because* they are means toward the attainment of their goals (Nagel 1977). The relation between a goal and its means is change-involving and invariant in that if x_e is the system's goal and (under a set of conditions) x_c is the means to the attainment of x_e, then, given x_e as a goal, x_c will occur as a matter of regularity (under those conditions). Furthermore, an intervention on x_e—one that changes the goal—would bring about a difference in x_c (the means). But not just any difference; generally, the new value of x_c will be one that is conducive to the new goal, x_e.

A purposive phenomenon instantiates two distinct invariance relations: one with its mechanical causes and the other with its goal (Walsh 2007). Consequently, the occurrence of a goal event is robust across two distinct sets of counterfactual conditions. From the causal/mechanical point of view, it is robust across a set of conditions in which the mechanical causes are held constant. From the goal-directed point of view, the occurrence of the event is robust across a set of conditions in which the goal state is held constant. These dimensions of counterfactual robustness are orthogonal in an important way. Holding the causes constant ensures that the effect will happen *whether it is a goal or not* (That is Aristotle's point). Similarly, the outcome, *qua* goal, demonstrates a measure of independence of the particular causal details. Holding the goal state constant ensures that it would occur, whether it occurred by this particular set of causes or not (That is Aristotle's point too). Mechanisms and goals underwrite distinct, orthogonal invariance relations.

Mechanist and purposive invariance relations have the same structure. If the former enter into genuine explanations, the latter should too. Just as there are mechanical explanations that identify the change-involving relation between a mechanism and its effect, there should also be purposive explanations that identify the change-involving invariance relation that holds between a goal and the means to its attainment. That, I take it, is one of the important lessons to be learned from Aristotle's doctrine of the four '*aitea*'.

2.3 Completeness and Complementarity

This insight must be allied to a lesson learned from modern mechanism—namely that invariance alone is insufficient for an explanation. There must also be a description that elucidates the explanatory nature of the invariance relation. Indeed, on Aristotle's approach to explanation, mechanistic and purposive explanations are marked out by different kinds of descriptions. In distinguishing 'material/efficient' cause explanations (those I am calling 'mechanistic') from 'formal/final' cause explanations (those I am calling 'purposive'), Aristotle implicitly appeals to the description relativity of explanations.

This is most clearly evident in the biological works. We find the features of organisms explained *both* by appeal to their efficient/material causes and by appeal to their formal/final causes. Each of these explanations involves a different invariance relation *and* a different kind of description. For example, respiration exchanges cool external air for warm, internal air. It happens *because* the heated lungs expand, drawing cool, external air in. Heat is exchanged between the hot internal organs and the cool air. Thus cooled, the lungs contract, expelling the warmed air (Johnson 2005). This is the efficient cause (mechanistic) explanation; its bottom out activities include expansion, contraction, inhalation and expulsion. Respiration also occurs *because* the overheated internal organs must be cooled in order to maintain their proper functioning. The effect of exchanging gases in the lung is the cooling of the internal organs. This is the formal/final explanation. So we explain respiration in two ways: (1) by appeal to its mechanical causes under a description that reveals the productive relation between respiration and its mechanisms and (2) by appeal to its effects under a description that identifies the contribution of respiration to the goals of the organism. The explanandum—respiration—instantiates two relevant invariance relations: one with its mechanisms (the expansion and contraction of tissues with heat) and the other to its goals (the cooling of internal tissues). The relevant description applied to each 'illuminates' the explanatory nature of the relation. This explanation 'twice over' runs throughout Aristotle's biology: 'there are very many things of this sort, especially among things which are constituted by nature are being so constituted; for nature makes them, on the one hand for the sake of something, and on the other out of necessity' (*Post* ii 11, 94b 34–37) (quoted in Johnson 2005: 58).

Formal/final and material/efficient cause explanations are each complete, and they are mutually complementary. The mechanical description completely specifies how the phenomenon is produced and the purposive description completely reveals the way in which the phenomenon contributes to the attainment of the organism's goal. They complement each other in that each elucidates a feature of the phenomenon that the other does not. Frank Lewis echoes this point in his account of the relation between Aristotle's formal/final and material/efficient explanations:

> … these two sets of causes offer differing but complementary explanations of the *same* things. Although the different explanations by themselves give only part of the full explanatory story, each can be seen on its own terms complete. (Lewis 1988: 61 emphasis in original)

The completeness and complementarity of mechanistic and purposive explanations, however, are not sufficient to establish them as discrete explanatory modes. The reason is that two mechanistic explanations can be complete and complementary, even if one can be reduced to the other. The completeness of an explanation is simply what modern mechanists call 'descriptive adequacy'. Descriptive adequacy, in turn, is relative to a particular set of 'bottom out' activities manifested at a given level of organisation. To return to our muscle contraction example, myosin fibres 'pull' and 'twist' the actin fibres. Pulling and twisting are bottom out activities at the macrostructural level of organisation. Yet the 'pulling' and 'twisting' of actin and myosin have complete chemical explanations too, whose bottom out activities include depolarization, hydrolysis, etc. These explanations are complementary in that each accentuates a different feature of the relation between the structure of muscles and their ability to contract.[10] Nevertheless, the information provided by the coarse-rain description is entailed by the information provided in the fine-grain description. The micro-level descriptions of these activities can replace the macro-level descriptions *without explanatory loss*. It is because of this replacement without explanatory loss that we say that the macro-level explanation can be reduced to the micro-level one. So, functional-morphological and microchemical explanations of muscle contraction are complete and complementary, but they are not distinct modes of explanation in the way that Aristotle supposes purposive and mechanistic explanations to be.

2.4 Autonomy

In order for mechanistic and purposive explanations to be distinct explanatory modes, they must also be mutually autonomous. One explanation, $E_{1,}$ is autonomous of another, E_2, if each completely explains the same phenomenon, and yet E_2 cannot

[10] Cf. Jackson and Pettit: 'Explanations of different causal grain are complementary' (2004: 178).

replace E_1 without explanatory loss.[11] Aristotle's respiration example illustrates this kind of mutual autonomy. There is nothing about the description that illuminates the productive mechanisms of respiration that also describes what it is for (for all the mechanistic explanation says, it might be the function of respiration to exchange CO_2 for oxygen!). Conversely, there is nothing about the goal that respiration subserves that informs us the mechanism by which it operates (for all this explanation says, air might be drawn into (mammalian) lungs by the action of the diaphragm and the intercostal muscles).

More schematically, the point is this. A goal-directed system has a set of goal-tropic trajectories, T, and a set of conditions, C, such that trajectory, t_i ($t_i \in T$), occurs under conditions, c_i ($c_i \in C$). Were some other condition, c_j ($c_j \in C$), to occur then some other trajectory, t_j ($t_j \in T$), would also. The set of ordered pairs of conditions and trajectories $\{<c_i, t_i>, \ldots, <c_k, t_k>\}$ that describes the goal-directed behaviour of the system instantiates two distinct invariance relations, one with the mechanism that produces t_i in c_i and the other with the goal, g, that the t_is all realise. Each of these relations, in turn, is elucidated by a different description. The mechanistic description describes how the mechanism produces t_i given c_i. The emergent teleological description describes how t_i conduces to g. The description that elucidates the way in which t_i conduces to the fulfilment of g does not elucidate the way in which the mechanism produces t_i (and vice versa). Consequently, we cannot substitute the description relevant to one invariance relation for the description relevant to the other without explanatory loss.

That is not quite sufficient to establish mutual autonomy. If a purposive explanation appeals to the invariance relation between the goal-tropic behaviours of the system $\{<c_i, t_i>, \ldots, <c_k, t_k>\}$ and the goal, then, on the principle that every phenomenon has a mechanistic explanation, this must have one too. So, every purposive explanation has a mechanistic counterpart. A purposive explanation is autonomous only if it cannot be superseded by its mechanistic counterpart.

The mechanistic explanation of the fact that every trajectory in T realises goal, g, proceeds by identifying the mechanism that causes each trajectory to end in g. The relevant description simply elucidates the way in which trajectory, t_i, produces g (given conditions c_i). But it does not cite the fact that the g is a goal or that the end-points are relevantly similar. One reason for this is that the mechanical invariance relation holds equally over *both* those trajectories that end in g and those that do not. In effect, the mechanistic explanation of why a trajectory ends in g is no different in kind from the explanation of why *another* trajectory of the system ends in a non-g state. Whether a particular trajectory ends in a goal state or not is strictly incidental to its mechanistic explanation. Not so for the purposive explanation. It describes the invariance relation between the trajectories in T and the goal, g, just as the mechanistic explanation does. But the relevant description cites the fact that g is a goal of

[11] I take it that explanatory autonomy is non-symmetrical. Typically, in a mechanistic reduction the reducing explanation replaces the reduced explanation without explanatory loss but the converse relation does not hold.

the system and, in any condition c_i ($c_i \in C$), the system has the capacity to produce a trajectory that realises g (and that t_i realises g in c_i). The fact that the endpoint, g, is the same for all the t_is and is a goal, is a crucially important part of the explanatory description.

The important differences between the mechanistic and purposive explanations of the dynamics of goal-directed systems are the following: (1) *the mechanism that produces the t_is appears essentially in the description of the relation between t_i and g in the mechanical explanation, but not in the purposive explanation*, and (2) *the fact that g is a goal appears essentially in the description of the relation between t_i and g in the purposive explanation, but not in the mechanical explanation.*[12] We cannot replace the mechanistic explanatory description for the purposive description (or vice versa) without explanatory loss. Thanks to description relativity, mechanical and purposive explanations are mutually autonomous explanatory modes.

Goals explain in a way that mechanisms do not. Having a goal is an emergent property of a system, in the trivial sense that goal-directed systems have goals whether or not their parts do. So, purposive explanations are emergent explanations. (I shall call this approach to purposive explanations 'emergent teleology'.) Aristotelian explanation, then, provides a model for explanatory emergentism. Explanatory emergentism is the view that every phenomenon has a complete mechanistic explanation, while some phenomena also have complete, autonomous emergent explanations. These appeal to emergent properties of complex systems. In the case of emergent teleology, they appeal to the goals of a goal-directed system.[13]

2.5 Downward Explanation

We saw that attempts to cast emergence as a robust metaphysical doctrine run aground on their commitment to reflexive downward causation. The problem with reflexive downward causation is that it implausibly, it requires that the properties of a complex entity have causal autonomy over the capacities of the parts. Teleological emergentism incurs an analogous commitment. It holds that the capacities of a system as a whole *explain* the activities of its parts. Call this 'reflexive downward explanation'. If reflexive downward explanation requires reflexive downward causation, then explanatory emergence is no better off than its metaphysical counterpart.

[12] This distinction between mechanistic and purposive explanations is reminiscent of Bedau's (1998) distinction between grade 2 and grade 3 teleology. Only in grade 3 (genuine) teleology does the fact that the goal state is a goal appears in the scope of the explanans. Unlike Bedau, however, I do not think that the concept of a goal is inherently normative (2008).

[13] I leave open the question whether there are other forms of emergent explanations, although I'm inclined to believe that the argument can be extended to show that there are emergent statistical explanations too.

There are two structural requirements for a goal-directed system: causal repertoire and downward regulation. Each part of a system has a causal repertoire, a range of activities it can engage in. Repertoire is important; it allows a goal-directed system to adopt any of a range of goal-tropic trajectories, according to the circumstances. But it also has the consequence that the system has the capacity, on any occasion, to produce a (usually much larger) range of non-goal-tropic trajectories. It is the regulatory architecture of the system as a whole that preferentially induces the parts to undertake those activities that realise goal-tropic trajectories (Kauffman 1993). The regulatory architecture of the system as a whole exerts a causal influence on the activities of the parts. This relation between the architecture of the system and the activities of the parts I call 'reflexive downward regulation'.[14] It is because a goal-directed system has this capacity for reflexive downward regulation that we can appeal to the goal of a system to explain the activities of the parts. Reflexive downward explanation requires reflexive downward regulation.

Reflexive downward regulation in no way suffers from the incoherence of reflexive downward causation. Reflexive downward causation requires that some emergent property causes the parts of a system to take on a causal capacity *that would not otherwise be in their causal repertoire*. Reflexive downward regulation does not confer on a part causal powers it does not otherwise have. It simply introduces a bias in favour of some of the activities in a part's repertoire over others. Because reflexive downward explanation requires only reflexive downward regulation—and not reflexive downward causation—emergent properties of a complex system can have explanatory autonomy over the actions of their parts, even if they have no causal autonomy.

3 Miscibility

One of the principal objections to emergent explanations is that they are otiose. Kim (1989), for example, argues that the emergent properties of complex entities play no autonomous explanatory role because every phenomenon has a complete mechanistic explanation that appeals to its parts. The supposition here is that emergent explanations are redundant because there is no unexplained residue left over after mechanistic explanation has done its work. Kim is relying here on the principle of explanatory exclusion:

> Roughly, the principle says this: No event can be given more than one *complete* and *independent* explanation. (Kim 1989: 79)

[14] P.W. Anderson's (1972) famous appeal for emergence is made on the grounds of what I am calling 'reflexive downward regulation'.

It is worth noting that traditional approaches to emergence also presuppose explanatory exclusion. They search for a domain of phenomena that cannot be given strictly mechanistic explanations, in order to carve out a role for emergence. But this search is futile; every phenomenon has a complete mechanistic explanation. The completeness of mechanism and explanatory exclusion together more or less entail that there are no genuine non-mechanical explanations. The completeness of mechanism is unassailable, so the emergentist, I maintain, should deny explanatory exclusion.

Explanatory emergentism posits a special relation between mechanistic and emergent explanation I call 'miscibility', a term borrowed from analytic chemistry. Two substances, for example water and ethanol, are miscible if they can be mixed, one with the other, without remainder (or residue). Two substances, for example, oil and water, are immiscible if they cannot be mixed. Where substances are immiscible, we find a boundary layer between them, on either side of which one substance excludes the other. Where substances are miscible, we do not find a boundary layer; both substances coexist in every region of the mixture. Mechanistic and emergent teleological explanations are miscible in the sense that where both apply, they do so over a single domain of phenomena, not over disjoint domains. There is no boundary between the phenomena to which emergent teleological explanations apply exclusively and the phenomena to which mechanical explanations apply. There are no gaps in the domain of phenomena over which mechanistic explanations apply to be filled by emergent teleological explanations.

Miscibility, I contend, offers a plausible alternative to explanatory exclusion; a given phenomenon *can* have more than one complete explanation. Mechanist and purposive explanations are complete, complementary and autonomous. Each is complete in the sense that it is descriptively adequate. Each identifies and illuminates a different feature of the system: The mechanistic explanation describes how the system produces its effects and the purposive explanation describes how those effects conduce to the fulfilment of the system's goals. They are autonomous in the sense that neither explanatory description can be substituted for the other without explanatory loss.

To recap, developing a robust emergentism requires two things. The first is to cast emergence as an explanatory—rather than a metaphysical—thesis. Explanatory emergence is the view that the emergent properties of (some) complex entities have explanatory autonomy over the activities of the parts, even if they have no causal autonomy. The second is to displace the presumption of explanatory exclusion. From the fact that every phenomenon has a mechanistic explanation, it does not follow that there is no role for emergent explanations. A single phenomenon may be susceptible of multiple explanatory descriptions, and hence, it may be subject to multiple autonomous explanations.

It remains to be seen whether there genuinely is a need in the life sciences for emergent teleological explanations in addition to mechanistic ones. Here, I think the organocentric program of evo-devo offers a hint. The capacities of organisms figure in genuinely emergent teleological explanations, and these are indispensable to an understanding of adaptive evolution.

4 The Autonomy of Evo-Devo

Organisms are the very paradigm of goal-directedness: 'you cannot even think of an organism … without taking into account what variously and rather loosely is called adaptiveness, purposiveness, goal seeking and the like' (Von Bertalanffy 1969: 45). It is clear that organisms' various goals figure unproblematically in emergent teleological explanations (Ayala 1970). For example, the mammalian thermoregulatory systems constitute a goal-directed system. Its goal is the maintenance of the proper temperature of the organism—that temperature that is most conducive to the organism's goal of survival. We explain why a particular episode of, say, vasodilation occurs in the skin by citing the fact that vasodilation helps to dissipate heat by increasing blood flow to the outer surface of the organism. This is an unexceptionable example of an emergent teleological explanation. It is descriptively adequate. It has a mechanical counterpart with which it is miscible and complementary, but neither replaces nor reduces the other. The question for evo-devo is whether the goal-directed capacities of organisms figure irreducibly in the explanation of adaptive evolution.

4.1 *Two Conceptions of Adaptive Evolution*

The principal objective of evolutionary biology is to explain the distinctive bias in biological form and function. The bias consists in the fact that certain traits occur regularly in organisms precisely because they are conducive to survival and reproduction. This bias is correctly attributable to the process of evolution. But, there are two ways of conceiving of that process. One is orthodox, well established and enshrined in the modern synthesis interpretation of evolution. The other is radically different and, I believe, implicit in the evo-devo approach.

The modern synthesis conception is grounded in a commitment to sub-organismal mechanism. It locates the source of adaptive evolution in the actions of genes/replicators and proceeds by elucidating the activities of genes/replicators in the production of phenotypes, both novel and recurrent. Phenotypes vary in their contribution to organismal survival and reproduction according to the replicators that produce them. Consequently, some genes/replicators systematically leave more of their copies in future generations than others. In this way, populations change in their genetic/replicator structure, and *concomitantly* they change in the phenotypic character of the organisms they comprise: hence the bias in form and function. The ultimate source of evolutionary novelties, on this view, is random mutation. Adaptive evolution proceeds by the gradual accretion (and recombination) of very small, lucky mutations.

There are three salient features of the modern synthesis conception of adaptive evolution that are worth special attention. The first is that the robust correlation between genotype and phenotype is taken to be a primitive feature of the activities of genes. The second is that it accords no ineliminable role to the capacities of organisms. The features of organisms appear only as explanandum in this explanatory scheme. The third is that it accords an ineluctable explanatory role to chance.

The evo-devo conception of adaptive evolution is distinctive. The fundamental explanatory principle is not the stereotypical activities of genes or the randomness of mutation rather; it is the lability of organisms. Organisms ensure their own survival despite the vicissitudes of internal and external conditions by effecting compensatory changes during their development (Kitano 2004). This capacity is called 'plasticity'; it is the very nature of an organism:

> the organism is not robust because it is built in such a manner that it does not buckle under stress. Its robustness stems from a physiology that is adaptive. It stays the same, not because it cannot change but because it compensates for change around it. The secret of the phenotype is dynamic restoration.... (Kirschner and Gerhart 2005: 108–109)

Plasticity confers on organisms, and their parts, an enormous repertoire of forms and activities. The plastic response of organisms during ontogeny serves to direct development toward the reliable production of a viable organism. A novel environmental or developmental circumstance requires the organism to make compensatory changes that secure the survival of the under these new conditions. This is known as phenotypic accommodation:

> Phenotypic accommodation is adaptive mutual adjustment, without genetic change, among variable aspects of the phenotype, following a novel or unusual input during development. (West-Eberhard 2003: 98)

The result of accommodation is often a novel adaptive phenotype. *These features occur precisely because they are conducive to the survival of the organism.*

Because each of the parts of a developmental system has a huge phenotypic repertoire, there are many developmental mechanisms capable of producing any given phenotype. So where a phenotype is recurrent in a population, there will be an enormous range of developmental systems within the population that have the potential to produce it. The plasticity of the organism as a whole regulates the activities of its developmental systems toward the production of these viable phenotypes. Thus, the origination and spread of adaptive novelties requires no mutation. It simply requires the plasticity of organisms (West-Eberhard 2005a).

Similarly, the developmental entrenchment of a novel variant in a population is driven by organismal plasticity. Each gene or gene system has the latent capacity to contribute to the production of a huge array of phenotypes. Changing from one productive role to another often involves only very minor alterations in the gene or gene system's regulatory relations (von Dassow and Munro 1999). Genetic resources that produce one phenotype in one regulatory context are co-opted to produce another phenotype in other contexts (Wray 2007; True and Carroll 2002). In this way, existing developmental structures are recruited into the reliable production of novel phenotypes. This can occur through very minor mutations in regulatory genes or no mutation at all:

> … the evolution of organismal form is much less a direct consequence of mutational genetic innovation, as believed earlier, but rather depends on continuing shifts, recruitments and

> re-wiring of regulatory interactions in development. Evolution seems to favour the generation of alternative genetic circuits which are subsequently co-opted-into new regulatory functions. (Muller 2008: 14)
>
> [A]ccommodation involves the re-use of old pieces in new places. (West-Eberhard 2005a: 617)

As a consequence, the developmental mechanism that produces an adaptive phenotype becomes progressively routinised (entrenched) (Schmalhausen 1949). The suggestion here is that the tight correlation between genotype and phenotype is a highly derived feature, not a primitive property of the activities of genes (Newman et al. 2006; Newman and Muller 2007; Salazar-Ciudad et al. 2001). The genotype/phenotype correlation is the consequence of a significant amount of 'reflexive downward regulation' of gene function by the plasticity of organisms.

This cursory overview of salient features of modern synthesis evolutionary biology and evo-devo should make vivid the differences between them. As I make it out, there are three important ones. The first is that the strong correlation between genotype and phenotype is a derived feature of both organisms and lineages. Genetic resources are co-opted by the organism to serve the purposes of securing the robust, recurrent production of novel phenotypes. The second is that change in the genetic structure of a population is generally the consequence of the adaptive bias of biological form, and not the other way around: 'genes are probably more often followers than leaders in evolutionary change' (West-Eberhard 2005b: 6543). The third is that the organism-centred approach leaves no ineluctable explanatory role to chance. Adaptive novelties appear in a population not ultimately because of random mutation but because of the adaptive plasticity of organisms. Even if genetic mutations are random occurrences, their phenotypic consequences are not. 'Variation is both less lethal and more appropriate to selective conditions than would be variation from random change' (Kirschner and Gerhart 2005: 221). The adaptive plasticity of organisms is pivotal to explaining these crucial features of adaptive evolution. Were it not for the plasticity of organisms, evolution would not be adaptive:

> Without developmental plasticity, the bare genes and the impositions of the environment would have no effect and no importance for evolution. (West-Eberhard 2005a: 6544)

4.2 *Emergent Explanation in Evo-Devo*

Evo-devo explains the bias in biological form and function by appeal to an emergent property of organisms—plasticity. In doing so it exposes a class of crucial invariance relations between an organism's goals of survival and processes of development. Plasticity is a goal-directed capacity, the ability to bring about changes in form *because* they are conducive to survival. These purposive invariance relations are *not incidental* to the process of adaptive evolution. Unless organisms had the capacity regularly to produce novel phenotypes and entrench them, adaptive evolution would not happen. The way in which plasticity contributes to survival, by producing novel adaptive phenotypes, is indispensable to the explanation of adaptive evolution. A complete account of adaptive evolution, then, requires us to acknowledge a class of emergent teleological explanations, those that appeal to the goal-directed capacities of organisms.

This explanatory dimension is missing from the sub-organismal, modern synthesis approach to explaining evolution. Of course every occurrence of an adaptive novelty will have a complete mechanistic explanation. But, the mechanistic explanation does not register the fact that an adaptive novelty occurs *because* it is conducive to the goals of survival and reproduction. It is *not* just a matter of chance whether a novel trait contributes to survival and reproduction. Of course it is one of the cornerstones of the modern synthesis that adaptive novelties arise ultimately as a matter of chance:

> The initial elementary events which open the way to evolution … are microscopic, fortuitous, and utterly without relation to whatever may be their effects upon … functioning. (Monod 1971: 118)[15]

This looks like an ineliminable metaphysical commitment to chance in adaptive evolution.

But a lesson recently learned from Aristotle becomes particularly germane here: mechanism cannot discern purposive regularities. Where goals are involved, mechanism cannot differentiate those events that occur by chance from those that occur because they fulfil a goal. What look like mere congeries of chance events from the mechanistic perspective may turn out to be a robust regularity from the perspective of purpose. It may well be, then, that the modern synthesis commitment to the ineluctable role of chance in evolution is less a deep metaphysical commitment than a superficial methodological artefact. I believe that it *is* methodological artefact—the unfortunate consequence of an overreliance on sub-organismal mechanism.

Newman et al. (2006) draw a distinction between what they call 'contingency' and 'inherency':

> Something is contingent if its occurrence depends on the presence of unusual … conditions that occur accidentally, conditions that involve a large component of chance, … something is inherent either if it will always happen … or if the potentiality for it always exists.[16]

The capacity for adaptive evolution is *inherent* in the plasticity of individual organisms. The inherency of adaptive evolution is exposed only once we countenance a class of autonomous, emergent explanations that appeal to the goal-directed capacities of organisms. The autonomy of evo-devo resides in the fact that it accords to the capacities of organisms an indispensable role in the explanation of adaptive evolution.

---

[15] It is interesting in this regard that Monod takes his title *Chance and Necessity* (and his inspiration) from the atomist, Democritus.

[16] The original source for the quotation is Eckstein (1980).

5 Conclusion

A complete understanding of the process of adaptive evolution requires us to acknowledge that biological form is biased by the capacity of organisms to originate and fix adaptive traits *because* such traits are conducive to survival. That in turn requires us to deploy a mode of emergent teleological explanation in biology. The single-minded pursuit of mechanism in modern biology has obscured the need for this kind of explanation. The great promise of evo-devo is that it points toward the ways in which the distinctive capacities of organisms contribute to the process of adaptive evolution. But the evo-devo programme cannot be brought to fruition unless its spectacular empirical advances are accompanied by a simple change in methodology. Evo-devo should renounce modern biology's exclusive reliance on mechanism in favour of explanatory emergentism. Only if it does so can it carve out an autonomous explanatory role for the capacities of organisms in evolution.

Acknowledgements I wish to thank audiences in Calgary, Edmonton, Vienna and Paris. In particular, I wish to thank Marc Ereshefsky, Fermin Fulda, Jesse Hendrikse, Philippe Huneman Gerd Muller, and Jacob Stegenga for helpful discussion. The bulk of this chapter was written while I was a visiting fellow at the Konrad Lorenz Institute. I thank all at the KLI for their marvellous hospitality.

References

Alexander, S. 1920. *Space, time and deity*. London: Macmillan.

Allen, G.E. 2005. Mechanism, vitalism and organicism in late nineteenth and twentieth-century biology: The importance of historical context. *Studies in the History and Philosophy of Biology and the Biomedical Sciences* 36: 261–283.

Anderson, P.W. 1972. More is different. *Science* 177: 393–396.

Aristotle. 2007. *Physics* Trans. R.P Hardie and R.K. Gray. http://classics.mit.edu/Aristotle/physics.2.ii.html

Ayala, F. 1970. Teleological explanations in evolutionary biology. *Philosophy of Science* 37: 1–15.

Bechtel, W. 2007. Biological mechanism: Organized to maintain autonomy. In *Systems biology*, ed. F.C. Boogerd, F.J. Bruggeman, J.-H.S. Hofmeyer, and H.V. Westerhoff, 269–302. Amsterdam: Elsevier B.V.

Bechtel, W., and A. Abrahamsen. 2005. Explanation: A mechanist alternative. *Studies in the History and Philosophy of Biology and Biomedical Sciences* 36: 421–441.

Bechtel, W., and R. Richardson. 1993. *Discovering complexity: Decomposition and localization as strategies in scientific research*. Princeton: Princeton University Press.

Bedau, M. 1997. Weak emergence. In *Philosophical perspectives 11: Mind, causation, and world*, ed. J. Tomberlin, 375–399. London: Blackwell.

Bedau, M. 1998. Where's the good in teleology? Repr. In *Nature's purposes: Analyses of function and design in biology,* ed. C. Allen, M. Bekoff, and G. Lauder, 261–291. Cambridge, MA: MIT Press.

Bedau, M. 2008. Is weak emergence just in the mind? *Minds and Machines* 18: 443–459.

Broad, C.D. 1925. *Mind and its world*. London: Routledge and Kegan Paul.

Callebaut, W., G.B. Muller, and S.A. Newman. 2007. The organismic systems approach: Evo Devo and the streamlining of the naturalistic agenda. In *Integrating evolution and development: From theory to practice*, ed. R. Sansom and B. Brandon, 25–92. Cambridge, MA: MIT Press.

Cartwright, N. 2004. Causation: One word, many things. *Philosophy of Science* 71: 805–819.

Craver, C. 2007. *Explaining the brain: Mechanisms and the mosaic unity of neuroscience*. Oxford: Oxford University Press.

Davidson, D. 1967. Causal relations. *Journal of Philosophy* 64: 691–703.

Dawkins, R. 1982. *The selfish gene*. Oxford: Oxford University Press.

Dawkins, R. 1998. Universal Darwinism. In *The philosophy of biology*, ed. M. Ruse and D.L. Hull, 15–37. Oxford: Oxford University Press.

Eckstein, H. 1980. Theoretical approaches to explaining violence. In *Handbook of political conflict*, ed. T.R. Gurr, 135–167. New York: Free Press.

Gilbert, S.F., and S. Sarkar. 2000. Embracing complexity: Organicism for the 21st century. *Developmental Dynamics* 219: 1–9.

Glennan, S. 1996. Mechanisms and the nature of causation. *Erkenntnis* 44: 49–71.

Glennan, S. 2002. Rethinking mechanistic explanation. *Philosophy of Science* 64: 605–626.

Goodwin, B. 1994. *How the leopard changed its spots: The evolution of complexity*. London: Touchstone Press.

Hankinson, J. 2003. *Cause and explanation in ancient Greek thought*. Oxford: Oxford University Press.

Humphreys, P. 1997. How properties emerge. *Philosophy of Science* 64: S337–S345.

Huneman, P., and P. Humphreys. 2008. Dynamical emergence and computation: An introduction. *Minds and Machines* 18: 425–430.

Illingworth, J.A. 2008. *Muscle structure and function*. www.bmb.leeds.ac.uk/illingworth/muscle/index.htm

Jackson, F., and P. Pettit. 2004. In defence of explanatory ecumenism. Repr. In *Mind, morality and explanation,* ed. F. Jackson, P. Pettit, and M. Smith, 163–185. Oxford: Oxford University Press, 2004.

Johnson, M.R. 2005. *Aristotle on teleology*. Oxford: Oxford University Press.

Kauffman, S. 1970. Articulation of parts explanation in biology and the rational search for them. In *Boston studies in the philosophy of science*, vol. 8, eds. R.C. Buck and R.S Cohen, 257–272. Boston: Springer.

Kauffman, S.A. 1993. *The origins of order*. Oxford: Oxford University Press.

Kauffman, S.A. 1995. *At home in the universe*. Oxford: Oxford University Press.

Kim, J. 1989. Mechanism, purpose and explanatory exclusion. In *Philosophical perspectives 3: Philosophy of mind and action theory*, ed. J. Tomberlin, 77–108. Atascadero: Ridgeview Publishing Company.

Kim, J. 2006. Emergence: Core ideas and issues. *Synthese* 151: 547–559.

Kirschner, M., and J. Gerhart. 2005. *The plausibility of life: Resolving Darwin's dilemma*. New Haven: Yale University Press.

Kitano, H. 2004. Biological robustness. *Nature Reviews Genetics* 5: 826–837.

Lewis, F.A. 1988. Teleology and material/efficient causes in Aristotle. *Pacific Philosophical Quarterly* 69: 54–98.

Machamer, P., L. Darden, and C. Craver. 2000. Thinking about mechanisms. *Philosophy of Science* 57: 1–25.

Mill, J.S. 1843. *A system of logic*. London: Longmans.

Monod, J. 1971. *Chance and Necessity*. Trans. A. Wainhouse London: Penguin.

Muller, G.B. 2007. Evo-devo: Extending the evolutionary synthesis. *Nature Reviews Genetics* 8: 943–949.

Muller, G.B. 2008. Evo-devo as a discipline. In *Evolving pathways: Key themes in evolutionary developmental biology*, ed. A. Minelli and G. Fusco, 3–29. Cambridge: Cambridge University Press.

Nagel, T. 1977. Teleology revisited. *Journal of Philosophy* 76: 261–301.

Newman, S.A., and G.B. Muller. 2007. Genes and form. In *Genes in development: Re-reading the molecular paradigm*, ed. E. Neuman-Held and C. Rehman-Suter. Durham: Duke University Press.

Newman, S.A., G. Forgacs, and G.B. Muller. 2006. Before programs: The physical origination of multi-cellular forms. *International Journal of Developmental Biology* 50: 289–299.
O'Connor, T. 1994. Emergent properties. *American Philosophical Quarterly* 31: 91–104.
O'Connor, T. 2000. Causality, mind and free will. *Philosophical Perspectives* 14: 105–117.
Oyama, S. 1985. *The ontogeny of information*. Raleigh.: Duke University Press.
Richardson, R., and A. Stephan. 2007. Mechanism and mechanical explanation in systems biology. In *Systems biology*, ed. F.C. Boogerd, F.J. Bruggeman, J.-H.S. Hofmeyer, and H.V. Westerhoff, 123–144. Amsterdam: Elsevier B.V.
Rosenberg, A. 2006. *Darwinian reductionism: Or how to stop worrying and love molecular biology*. Chicago: University of Chicago Press.
Rueger, A. 2000. Robust supervenience and emergence. *Philosophy of Science* 67: 466–489.
Salazar-Ciudad, I.S., S.A. Newman, and R.V. Sole. 2001. Phenotypic and dynamical transitions in model genetic networks. I. Emergence of patterns and genotype-phenotype relations. *Evolution and Development* 3: 84–94.
Salmon, W. 1984. *Explanation and the causal structure of the world*. Princeton: Princeton University Press.
Schmalhausen, I.I. 1949. *Factors of evolution*. London: University of Chicago Press (1986).
Silbersteen, M., and J. McGeever. 1999. The search for ontological emergence. *The Philosophical Quarterly* 49: 182–200.
Sober, E. 1999. The multiple realizability argument against reductionism. *Philosophy of Science* 66: 542–564.
Strevens, M. 2004. The causal and unification approaches to explanation unified—Causally. *Nous* 38: 154–176.
Symons, J. 2008. Computational models of emergent properties. *Minds and Machines* 18: 475–491.
True, J.R., and S.B. Carroll. 2002. Gene co-option in physiological and morphological evolution. *Annual Reviews of Cell and Developmental Biology* 18: 53–80.
Von Bertalanffy, L. 1969. *General systems theory*. New York: George Barziller.
Von Dassow, G., and E. Munro. 1999. Modularity in animal development and evolution: Elements of a conceptual framework for Evo-Devo. *Journal of Experimental Zoology (Mole Dev Evol)* 285: 307–325.
Walsh, D.M. 2007. Teleology. In *Oxford handbook of philosophy of biology*, ed. M. Ruse, 113–137. Oxford: Oxford University Press.
Waters, C.K. 1996. Why the antireductionist consensus won't survive the case of classical Mendelian genetics. Repr. In *Conceptual issues in evolutionary biology*. 2nd ed., ed. E. Sober, 401–418. Cambridge: MIT, 1998.
Waters, C.K. 2008. Beyond theoretical reduction and layer-lake anti-reductionism: How DNA retooled genetics and transformed biological practice. In *Oxford handbook of philosophy of biology*, ed. M. Ruse, 238–262. Oxford: Oxford University Press.
Webster, G., and B. Goodwin. 1996. *Form and transformation: Generative and relational principles in biology*. Cambridge: Cambridge University Press.
West-Eberhard, M.J. 2003. *Developmental plasticity and evolution*. Oxford: Oxford University Press.
West-Eberhard, M.J. 2005a. Phenotypic accommodation: Adaptive innovation due to developmental plasticity. *Journal of Experimental Zoology (Mole Dev Evo)* 304B: 610–618.
West-Eberhard, M.J. 2005b. Developmental plasticity and the origin of species differences. *PNAS* 102: 6543–6549.
Wimsatt, W. 2000. Emergence as non-aggregativity and the biases of reductionism. *Foundations of Science* 5: 269–297. Repr. In *Re-engineering philosophy of limited beings*, ed. W. Wimsatt, 274–312, 2007.
Woodward, J. 2002. What is a mechanism: A counterfactual account. *Philosophy of Science* 69: S366–S377.
Woodward, J. 2003. *Making things happen: A theory of causal explanation*. Oxford: Oxford University Press.
Wray, G.A. 2007. The evolutionary significance of *cis*-regulatory mutations. *Nature Reviews Genetics* 8: 206–216.

Does Oxygen Have a Function, or Where Should the Regress of Functional Ascriptions Stop in Biology?

Jean Gayon

Abstract Biologists apply the notion of function to almost every type of structure and process that enters into descriptions of biological phenomena. They can also generate alarmingly long regresses: x can be the function of y, which is the function of z ... and so on. But the functional regress must stop somewhere. This chapter investigates whether the philosophical theories restrict the regress of functional attributions by asking if they legitimate making such attributions to structures at elementary levels of organization (atoms and elementary molecules) and to structures at higher ones (organisms and species). First, I propose a classification of the current theories of functions into three categories, rather than the usual two: Larry Wright's "etilogical theory" is definetly something different from the "selective etiological theories" that have been developed after him. Then I examine whether these theories can admit or not the ascription of functions to very low or very hig levels of organization. At the most elementary levels, functional ascriptions are unacceptable for the selective etiological theory of functions, because atoms or elementary molecules are not units of selection; they are less problematic for the systemic theory of functions, and also for Wright's original "etiological theory," provided that the composition and behavior of the parts constituting the system involved are precisely stated. At the level of organisms and species, functional ascriptions are possible within both the selective etiological and the systemic theory, but this will heavily depend on the theoretical framework involved in both cases. These limit cases show that the selective conceptions of functions are less tolerant than the systemic ones. They also suggest, as already noted by William Wimsatt, that functions are more convincingly ascribed to processes than to structural entities.

J. Gayon (✉)
Institut d'histoire et de philosophie des sciences et des techniques,
Université Paris 1-Panthéon Sorbonne, Paris, France
e-mail: jean.gayon@gmail.com

P. Huneman (ed.), *Functions: selection and mechanisms*, Synthese Library,
Studies in Epistemology, Logic, Methodology, and Philosophy of Science 363,
DOI 10.1007/978-94-007-5304-4_4,

1 Introduction

Biologists apply the notion of function to almost every type of structure and process that enters into descriptions of biological phenomena. Among the structures to which they attribute functions, one finds the following: bodily systems, the names of which usually designate their imputed function—the circulatory system, the immune system, etc.; organs themselves; cells; organic macromolecules; smaller organic molecules, such as sugars and amino acids; and simple inorganic molecules, including O_2 and H_2O. Functions are even attributed to atoms and their ionic forms. For instance, biologists refer to the function of the ferric ion in hemoglobin. Biologists also impute functions to structures at higher levels of biological organization. Individual organisms, populations, and species are sometimes spoken of as having functions in relation to larger ensembles such as colonies, biotic communities, and ecosystems. Processes are accorded functions, too. It thus seems normal to ask "what is the function of the Krebs cycle in aerobic organisms?", or "what is the function of paradoxical sleep in vertebrates?"

Not only are statements about functions omnipresent in life sciences, they can also generate alarmingly long regresses. In his *Critique of the Faculty of Judgment*, Immanuel Kant noted this problem. Although he never used the term "function," Kant emphasized that the relations among the parts of a living being can always be simultaneously understood as causal relations and as relations of means to ends. Once one identifies the end toward which something is a means, one can always inquire about the further end which that end serves. Yet, as Kant recognized, the instrumental regress must stop somewhere. Peter McLaughlin (2001) took Kant's remarks as the basis of his observation that while x can be the function of y, which is the function of z … and so on, the functional regress must stop somewhere.

Here, I will consider the extremely liberal attributions of functions in ordinary current biological practice under the light of the most widely discussed recent philosophical theories of function. I will investigate whether these philosophical theories restrict the regress of functional attributions by asking if they legitimate making such attributions to structures at elementary levels of organization (atoms and molecules) and to structures at higher ones (organisms and species).

2 Theories of Function: Three Families

First, let us recall the principal theories of function that have received philosophical attention since the 1970s. In a rigorous and profound work, Marie-Claude Lorne (2004) distinguishes three major families of such theories. The first comprises systemic theories, which Robert Cummins' theory (1975) exemplifies. It interprets biological functions as both relative to the biological system toward which they contribute and relative to a particular explanatory strategy. This strategy explains the capacity of a system by reference to the capacities of its parts. So according to Cummins' systemic theory, to say that something *I* has the function *F* in system *S* is attribute to *I* a capacity, the exercise of which plays a causal role in the emergence of a capacity of a larger system *S* of which *I* is part. For example, the contraction of the diaphragm determines the dilation of the pulmonary cavity in vertebrates with lungs. It does this by making the air pressure in the lungs fall, which causes air to rush in from outside. To say that the diaphragm's function is to dilate the pulmonary cavity is thus to say that its capacity to do so contributes causally to the emergence of the global capacity of the respiratory system—respiration.

The evolutionary history of a system is irrelevant to functional explanations given by the systemic theory. Such explanations consist only in analyzing a system and its component parts, in identifying the capacities of these parts, and in demonstrating how the exercise of these capacities contributes to the emergence of more complex capacities at higher levels of organization. Lorne astutely notes that understood in this way, functions are not properties that exist objectively or independently. Thus, on Cummins' original systemic conception, statements about functions are significant only in relation to an explanatory project, in this case, one which aims to make sense of the capacities of a hierarchically organized system.

The second major theoretical notion of function is Larry Wright's (1973). This concept, which Wright calls "etiological," is overtly teleological. According to Wright the function of *X* is *Z means*: (a) *X* is there because it does *Z* and (b) *Z* is a consequence (or result) of *X*'s being there. (b) expresses a dispositional clause; for instance, if *X* is an organ such as the kidney, the elimination of urea in our blood is the consequence of the presence of our kidneys. (a) is etiological in the sense that this clause explains "how *X* came to be there" (Wright 1973: 47); for instance, our kidneys are there (or have come to be there) because (among other things) they eliminate urea. Wright explains that natural selection is a plausible interpretation of this condition in current biology: "We can say that the natural function of something—say, an organ in an organism—is the reason the organ is there, by invoking natural selection. If an organ has been naturally differentially selected- by virtue of something it does, we can say the reason the organ is there is that it does that something" (Wright 1973: 46). From this quotation, it is clear that the subsequent "selective etiological theories" can be derived from Wright's theory, by just dropping the other condition (condition (b)) stated by Wright. This is what Neander explicitly stated in her 1983 Ph.D. (Neander 1983: 104–106). However, Wright rejected this option. I will return to this problem later, in my discussion on the function of oxygen. Furthermore, selection (either mental selection for artifacts or natural selection in the case of biological objects) is one possible background for Wright, not a necessary element of the theory.[1] Based in ordinary language analysis, Wright's etiological conception accentuates the irreducibility of the teleological aspect of all function attributions, whether they are made about biological things or about artifacts.[2]

[1] As shown by Lorne, this aspect of Wright's theory becomes clearer if one compares his 1973 article with his book on teleology (Wright 1976).

[2] In fact, Wright proposed his etiological conception of function in order to clarify problems in action theory, not biology. But I will not address that here.

Lorne resists other philosophers' tendency to credit Wright with the third family of theories of function, which she gathers under the name "selective etiological theories." Although these theories are related historically to Wright's "etiological theory," they are different because they closely link the notion of function to that of the evolution of biological systems by natural selection (whereas in Wright's conception, natural selection was only a possible background theory—see Wright 1976). The most widely held selective etiological theory is expressed as follows: "the function of a trait is the effect for which that trait was selected" (Neander 1991). According to this theory, to say that the function of the heart is to pump blood is to say that the heart exists because a structure similar and ancestral to the heart conferred advantage on the organisms that possessed it.

Lorne distinguishes three kinds of selective etiological theories. The first kind is Neander's standard account, which has already been presented. The second is Peter Godfrey Smith's "modern history theory of functions," which differs from the standard theory with respect to time. According to this concept, "functions are dispositions which explain the recent maintenance of a trait in a selective context" (Godfrey Smith 1994). What matters here is that the selective forces that have *maintained* a trait in the recent past can be different from the forces that were salient of the *origin* of the trait. As noted by Godfrey Smith, the modern history theory has something to do with Gould's and Vrba's notion of exaptation. Exaptations are characters that originated for one reason or another (selective or not) and that have been co-opted for new use. For instance, feathers in birds did not originate as adaptations for flight, because they appeared in animals that were not able to fly and were only co-opted for flight later. The third variant of etiological theories of function, according to Lorne, cancels any reference to the past. From this perspective, to attribute a function to a biological entity is to conjecture about that entity's contribution to fitness. In other words, it is to conjecture about its contribution to the survival and reproduction of the organisms that currently possess it. Lorne herself subscribes to this concept, which was originally defended in a systematic way by Bigelow and Pargetter, in their propensity view of functions: "something has a (biological) function just when it confers a survival-enhancing propensity on a creature that possesses it" (see Bigelow and Pargetter 1987 for this precise formulation in terms of propensity, but see also Ayala 1968 for the "contribution to fitness" contribution).

Lorne's classification of the theories of function conflicts with the classical backward-looking/forward-looking classification in two ways. First, Wright's original treatment of the etiological theory *is not* counted as a "selective etiological theory" (because it is not intrinsically selective). Secondly, the contribution to fitness' conception is counted as a member of the "selective etiological theories" family, instead of its not being backward looking.

Contemporary philosophical literature on function is much richer than this brief sketch can express (for a comprehensive view, see Allen et al. 1998; Ariew et al. 2002; Buller 1999; Lorne 2004; Gayon and de Ricqlès 2010). I will state at the end of this chapter why I favor this classification rather than the usual backward/forward. In the subsequent section, I use it as a convenient tool for raising the question: to which levels of OR organization is it reasonable to attribute functions in biology?

3 Functions and Levels of Organization

With rare exceptions, philosophers have been silent on the subject of function with respect to the levels of organization. The definitions of function that have been proposed refer to "traits" or "items." This is indeed a distinctive feature of modern philosophical debate on function (by "modern," I mean 1970 and after). Speaking of traits is a reasonable strategy for avoiding the traps of ordinary biological language, which tends to limit the range of things to which we can attribute functions by always using the term "function" in the context of "structure" and by applying this "structure/function" dyad only to organisms and their parts. The modern debate is not about "biological functions" in the ordinary sense that morphologists and physiologists give it (e.g., "the respiratory function," "the female reproductive function," and "the sensory function"). In this particular use of the word, i.e., this or that "function," what is at stake is, in fact, a combination of a certain structure and what this structure is supposed to do (the function *of* that structure). The modern debate is about ascribing a function to something ("function" *of* rather than "*a* function"). In such a context, attributing functions to "traits" covers any possible situation, at all possible levels of organization. Traits themselves can be structures, processes, or behaviors. Thus, philosophers have carefully produced definitions of function that are as general as possible. Quite often, modern philosophers who debate about functions aim to construct a notion that is broad enough to apply not only to living systems but also to artifacts (Vermaas and Houkes 2003) and in the context of cognitive science (Block 1980; Pacherie 1995).

I will remain within the biological realm here. In the rest of this chapter, I will draw attention to several levels of organization to which functional attributions could be problematic. My analysis should be taken as a preliminary step toward a more systematic analysis of how the modern philosophical debate over functional ascription could be articulated, with the traditional concept of biological function in terms of the structure/function debate (which remains largely dominant in the ordinary scientific practice). I will focus on three problematic cases: atoms and elementary molecules, organisms, and species. The reason for such a choice is that our common intuitions about functional ascriptions in biology are tailored to think about the function of this or that part or process in an organism. Therefore, one may conjecture that problematic instances will be more easily found either in the case of the most elementary parts or processes of an organism (those that are not distinctively biological) or in the case of supraorganismal entities (e.g., species) or else in the case of organisms themselves (which stand, most often, as the end point of functional ascriptions). In each case examined, I will rely on the classification of theories of function given earlier. On the whole, it will be seen that the etiological theories (especially the selective etiological theories) are more demanding than the systemic theories.

4 Can Elementary Molecules Have a Function?

Let us first consider the case of atoms and elementary molecules. What does it mean to attribute a function to a Fe^{+++} ion or to an O_2 molecule in an organism? Let us consider oxygen, one of the most abundant physical components of any organism. If we ask "what is the function of oxygen?" we are obviously not interested in the many molecules of which the oxygen atom is part in a living body. For instance, we are not interested in the four oxygen atoms included in a molecule of aspartic acid, 1 of the 20 amino acids found in the polypeptide chains composing proteins. Of course, due to the properties of the oxygen atom, these four oxygen atoms contribute to the overall properties of molecule aspartic acid, and to the biological roles of this molecule. But this is not the issue: we would hardly say that oxygen, as such, has a function or a biological role in the innumerable large or small molecules that contain atoms of oxygen and that obviously have a biological function in an organism. If biologists are to ascribe a biological role or function (I remain deliberately vague on that point) to oxygen, it is more likely, in cases where oxygen exists in a free state, able to have specific effects in a biological context as a molecule. The most obvious example can be found in the case of respiration. Oxygen is crucial to the energetic metabolism in every cell of every aerobic organism. Without it, cells would only be able to extract a tiny fraction of the energy that they actually extract from the breakdown of glucose and other organic molecules. The presence of free oxygen in the mitochondria of eukaryotes (or within the cytoplasm of aerobic prokaryotes) is necessary for the removal of electrons from certain coenzymes, which ultimately results in the creation of available energy. A network of complex molecules, called the electron transport chain, culls hydrogen ions (or protons) and their associated electrons from the coenzymes $NADH^+$ and $FADH_2$, ultimately shuttling them to half molecules of oxygen. This creates water, NAD and FAD, and energy. The electron transport chain is coupled with the Krebs cycle, which reduces NAD and FAD as part of its process of liberating energy from carbohydrates by breaking them down into CO_2. Clearly, oxygen plays an important biological role in this process. It plays the role of a combustive which, through a cascade of coupled reactions, makes it possible for cells to glean energy by breaking down carbohydrates into CO_2. In a way, oxygen fires up the carbohydrate-degrading machinery that makes energy available.

Let us interpret this in functional language. If oxygen plays a biological role, it would seem natural to say that it has a function. Note that this function is relative to a state of biological understanding. A century ago, one would not have said that the role of oxygen was to be the terminal electron acceptor in the mitochondria's respiratory chain. The question here is whether or not one has the grounds to speak of "function" in a case like this.

I will start from the selective etiological theories of function perspective. In Neander's backward approach, the function of a trait is equated with the effect for which it has been selected. Clearly, oxygen has not been *selected* for anything. While it makes sense to say that the proteins in the mitochondria's electron transport chain are the products of a process of natural selection, it seems nonsensical to say the same of oxygen. Even though oxygen molecules have invaluable biological effects, one cannot say that they have a function, at least not in the sense that the classical selective etiological theory gives. The fact that every nanosecond, huge numbers of oxygen molecules are in the right place in each of the mitochondria in each of the 60 trillion cells in a human body and that they perform a very specific action in the respiratory chain of each cell is a result of natural selection. But this result is a complex phenotype, which falls completely outside the class constituted by oxygen molecules. Thus, Neander's version of the selective etiological theory does not apply to oxygen. We certainly cannot paraphrase Neander and say, "the function of oxygen [in aerobic organism] is the effect for which oxygen was selected." The same could be said of the modern history theory of function. It makes no more sense to say that oxygen has been selected recently than to say it was selected in a remote past. What has been selected is a tremendously complex biochemical and morphological system that converts a very dangerous molecule into an essential resource for aerobic organisms, not oxygen as such. A similar argument also applies to the propensity interpretation of function, which also appeals (though prospectively, not retrospectively) to natural selection. We cannot say, paraphrasing Bigelow and Pargetter, that "oxygen has a (biological) function in aerobic organisms insofar as it confers a survival-enhancing propensity on a creature that possesses it." Of course, finding more or less oxygen in the environment may greatly affect the chances of survival and reproduction of a given organism. But the fact that there is more or less oxygen in the external milieu is not part of the organism's fitness. What might be part of its fitness is its behavioral ability to find places with an appropriate quantity of oxygen or to defend itself against either scarcity or excess of oxygen. For instance, mole rats are able to live in underground tunnels where the oxygen density is only 10% of what it is outside.

Now, if we adhere to the systemic theory instead of the selective etiological one, might we properly say that oxygen has a function?—definitely yes. The systemic theory defines something's function as its capacity, the exercise of which plays a causal role in the emergence of a more complex capacity of the system of which that thing is a part. Oxygen certainly plays a well-defined causal role in the elementary biochemical processes that define the respiratory chain in mitochondria (or in prokaryotic cells as a whole). One might object here that oxygen is not a constitutive and stable part of the mitochondria's (or the cells) respiratory system but rather an external affluent. However, this affluent is necessary for the system to perform its role. We should recall here that the systemic theory (also called causal role theory) of function is relative to a particular explanatory strategy. If the explanatory objective is to account for a metabolic process including a flux of molecules coming from the outside, these molecules are part of the so-defined dynamical system. In this respect, oxygen behaves like the photons captured by the rods and cones in our retinas. The presence of photons is a *sine qua non* of photoreception, but photons are not, properly speaking, a part of the visual system as currently described in anatomy, though they are part of a system involved in an explanation of vision. But, of course, without photons entering into it, there would be no vision. Since oxygen and photons contribute causally to the emergence of a global capacity (respiration and vision), they can be said to truly perform a function from the viewpoint of the systemic theory of function. From the physiological explanation viewpoint, they are part of the functioning of respiration or vision as currently described in literature.

The example of the molecule of oxygen provides us with an extremely crude contrast between the selective etiological theories and the systemic theories of function. In the former case, whatever version of the selective theory we chose, it is nonsensical to say that oxygen has a function. The only solution is to delegate the functional ascription to a rather complex set of traits that can truly be said to have been selected (or to contribute to the fitness of the organism). In the case of the systemic theory, it is nonproblematic to attribute a function to oxygen. Any description of the functioning of the respiratory chain in a standard contemporary textbook on physiology or molecular biology would offer a perfect example of the systemic theory. What has been said of oxygen could be said of a number of atoms or molecules (elementary or not) that intervene in metabolic processes, from the iron, potassium, or sodium atoms (or their ionic forms) to water and a number of simple molecules playing some causal role in the emergence of a biological capacity. In all cases, ascribing functions to them would be nonproblematic, while it would not be acceptable to do so for any version of the selective etiological theory.

So far, I have left Wright's etiological theory out of my discussion about the function of oxygen. Let us first note that Wright takes the example of oxygen and relies upon it in his 1973 article (Wright 1973: 159–161). I will not defend that Wright's original etiological conception could be a way of getting out of difficulties raised by the selective etiological theories here. I will just try to show why Wright's schema made it possible for him to state that oxygen has a function. In fact, the example of oxygen was a key step in the argument that led him to his well-known definition quoted above. Today, Wright's contribution is too often reinterpreted in the light of subsequent literature on the explicitly selective versions of the "etiological theory."

Oxygen arises twice in Wright's paper, in relation to two things done by this molecule in the course of the respiratory process, combining with hemoglobin and providing energy in oxidation reactions (Wright does not enter into the biochemical detail of the latter statement, but this is of no importance here). In both cases, Wright's first necessary condition for being a function or condition (a) ("*X* is there because it does *Z*")[3] is satisfied. According to current biological knowledge, it is true that oxygen *is there* (in our blood) because it combines with hemoglobin: "oxygen combines readily with hemoglobin, and that is the (etiological) reason it is found in human bloodstreams" (Wright 1973: 159). It is also true that "oxygen must be there (in the blood) because it produces energy" (Wright 1973: 159). But it would be nonsensical to maintain that the function of oxygen is to combine with hemoglobin. Therefore, there must be something different in the meanings attributed to the two words "because" in the statements above about what oxygen *does* (its various

[3] Note that in the 1976 book, the order of the two conditions (a) and (b) is reversed.

activities). Wright identifies that difference through the second condition, condition (b), of his definition of function "*Z* is a consequence (or result) of *X*'s being there." Whereas it makes no sense to say that the combining with hemoglobin is not a consequence of oxygen's being in our blood (but rather the reverse: oxygen is in our blood because of its combining with hemoglobin), we can definitely say, "producing energy *is* a result of [oxygen] being there" (Wright 1973: 161).

What is at stake in Wright's treatment of function is teleology. His account of functional ascriptions is a plea for something irreducibly teleological, expressed in condition (a). In sentences such as "oxygen is there *because* it produces energy," "kidneys are there *because* they eliminate metabolic wastes from the bloodstream," and "chlorophyll is there *because* chlorophyll enables plants to accomplish photosynthesis," the word "because" cannot be understood in the sense it has when we say that "oxygen is there in the blood because it combines with hemoglobin." Wright's conception of function, as Cummins' opposed conception, is epistemic, nonrealistic: it says something about the kind of inference that we make when we ascribe functions; neither Wright nor Cummins thinks that functions are something objective in nature. They are relative to our explanatory purpose. This is related to the nonessential role of natural selection in Wright's concept: natural selection is the only scientific *background* that Wright is able to mention for the application of his theory to biological objects, but natural selection is definitely not a part of his definition. This is why it is not a problem for Wright (no more than for Cummins) to ascribe functions to molecules (or to any part or process in a biological system). Within a certain theoretical context (where natural selection may play a key role), oxygen, although not selected as such, is part of a system designed by evolution, so that it makes it possible to say "oxygen is there because it provides energy in oxidation reactions."

I am not sure whether Wright did ever consciously raise the question of whether elementary molecules can have functions or not (probably not), but the case I have just been examining shows an important difference between his theory and current selective etiological theories. In her fascinating doctoral dissertation, which I already alluded to, Neander proposed dropping Wright's condition (b)[4] altogether and modifying condition (a) so as to have "*X* is there *because it was* selected because it does (results in) *Z*," so that the definition of function becomes "The function of *I* in *O* is to do *C* if *I* was selected (by natural selection) in *O because it does C*" (Neander 1983: 103, 107). In Neander's approach (as well as in all versions of etiological theories), ascribing functions means making (or aiming at making) an objective and realistic statement in the framework of a well-accepted scientific theory. Thus, in a sense, "functions" objectively capture a level of reality in nature, not only a level of explanation; consequently, one cannot ascribe functions to each thing that has a just "role" in living beings: oxygen does have a causal "role" (or perhaps many) in a number of organisms, but it has not a "function." The "selected effect" theory of function, if taken seriously, cannot be immune with respect to the issue of the units and levels of selection.

[4] Named "condition (2)," because Neander refers to the 1976 book, where Wright's two conditions are presented in reverse order.

5 Organisms and Above

Do we encounter similar problems when we attribute functions to entities at the highest levels of biological organization (organisms and above)? I will be briefer on these cases, but they raise problems similar to those developed above in the case of elementary molecules. In this more sketchy part of the chapter, I will concentrate on the two big families of theories of function considered today: the selective etiological theories and the systemic theories. I will no longer distinguish the variants of the selective etiological theories, because they are similar with respect to the problems examined. I will also put aside Larry Wright's original etiological theory.

Can we attribute functions to organisms *per se*? More precisely, can we attribute functions to organisms qua members of a certain type within a species? The question arises in several biological contexts: colonial organisms, social insects that segregate themselves into castes, and sexually reproducing species. In each case, members of different classes of organisms (e.g., castes and sexes) play different roles in larger ensembles such as colonies or species. Yet biologists usually attribute functions to an organism's *traits* (e.g., morphologies and behaviors), rather than to the organism itself. The organism itself is the system with respect to which biologists make *most* of their functional attributions. So far as it is the typical *terminus ad quem* of the regress of functions, the organism itself is not considered to be the proper object of functional attributions.

Nonetheless, attributing functions to organisms does not pose serious problems within the systemic theory of function. As long as organisms are components of the system in question, it is possible for them to have functions with respect to that system. Colonies provide examples of systems, with respect to which one could attribute functions to members of particular groups of organisms, such as castes of social insects. Species provide examples, too. As maximal panmictic communities, species are systems with respect to which one could attribute functions to male and to female organisms.

When it comes to selective etiological theories, the applicability of functions to organisms is less straightforward. According to standard evolutionary theory, selection maximizes the fitness only of individual organisms. Thus, a trait is selected only if it maximizes the probability that members of a particular class of organism will survive or reproduce. Yet within the framework of the selective etiological theory, in order for organisms as such to have functions, they would have to maximize the fitness of things, or classes of things, other than individual organisms. They might maximize the fitness of groups, entities that are more inclusive than individual organisms. Or, they might maximize inclusive fitness, which accrues to genes rather than the organisms that bear them. Thus, whether organisms can be accorded functions within the selective etiological theory depends upon the theoretical framework one uses to understand selection.

The last case I would like to mention is that of species. Can species *as such* have functions? The question arises in the context of ecology. One might say that different species can play the same role in an ecosystem or that one species can be replaced by another species that makes the same contribution to the maintenance of the ecosystem. The systemic theory has no trouble speaking of "function" in such a situation, because it makes no prescriptions about the nature of the systems within which an item can be said to have a function. For example, the systemic theory would have no trouble saying that the function of a certain animal species in a certain ecosystem is to disperse the seeds of a certain plant—as long as the causal role of this activity has been established.

In contrast, the selective etiological theory would have more difficulties with such assertions about the function of a species. Since ecosystems do not reproduce, the notion of selection does not apply to them, at least not in the ordinary sense of "selection." Of course, some ecological theories allow for a nontrivial notion of ecosystem selection, based upon persistence but not reproduction (Van Valen 1991; Blandin 2007; Bouchard 2008). This brings us to the same place that the question about the function of organisms did. The solution to the philosophical problem of whether species can have functions in the sense of the selective etiological theory depends on the kind of theory of selection one adopts.

6 Conclusion

What conclusions can we draw from the limited cases examined here? First, we can conclude that selective concepts of functions are the least tolerant toward these cases. Despite their apparent *naiveté* and intuitiveness, selective etiological theories (of all kinds) are more demanding than systemic ones. This is not surprising, because selective conceptions depend on a particular biological theory that can be instantiated by many different models. The systemic conception is more liberal, because it does not a priori assume specific theoretical commitments. It accommodates all of the limited cases examined here. Larry Wright's theory is also more liberal, for the very same reason. The less tolerant character of the selective theories of function is not a defect, but rather a quality. A major commitment of these theories of function is that they make the biological notion of function dependent upon a particular scientific theory, whereas the systemic theories and Wright's etiological-teleological theory consider it as a sort of universal conceptual tool for explanation in biology and technology. Because the selective theories of function depend upon the theory of natural selection, we can safely anticipate that current functional ascriptions might be seriously questioned, with respect to what it really means in terms of selective explanation. I have given only a few examples. Whether my diagnosis about the more demanding character of etiological theories can be extended or not will depend on a careful examination of a significant array of structures and processes all along the hierarchy of biological organization. This might be a good way of reconciling modern philosophical discussions about functions (discussions about functional ascriptions) with the more classical approach of biologists in terms of structures, processes, and functions.

The second conclusion of this chapter relates, precisely, to this tryad (structure, function, and function). Biologists do not attribute functions only to structures; they also attribute them to processes. Yet all of the limited cases examined here—molecules, organisms, and species—are structures. Taking account of processes would doubtlessly lead to a different understanding of the regress of functions.

William Wimsatt (2002) holds that, strictly speaking, one can only attribute functions to behaviors and operations, not to physical objects such as structures or systems. According to Wimsatt, vascular capillaries cannot have functions, but their contracting and dilating behavior can. Likewise, we cannot attribute a function to the heart, but we can attribute functions to its abilities to take in and pump out blood and to change the rate at which it beats. If one follows Wimsatt's recommendation, one would never speak as though the parts of organisms (or of more inclusive systems) had functions. One would only attribute functions to the behaviors they exhibit under given conditions. In Wimsatt's view, two objects are functionally equivalent if they do the same thing in comparable systems in similar environments. The physical objects, or structures, that perform these functions are not independent variables relative to the functions.

Endorsing Wimsatt's position would lead to a dynamic understanding of functions, a step in the direction of reconciling the notion of "function" with that of "functioning." From Wimsatt's position, we might also be able to steer around the problems with attributing functions to entities such as elementary molecules. While it may be problematic to attribute a biological function to oxygen if we adhere to the selective etiological theory of function, it is not so hard to attribute a function to the activity of capturing electrons and protons that oxygen performs and then to specify the relevant biological context within which this activity is performed, including some biological entities that could be truly said to have been "selected for." And then, perhpas, we might find some kind of compromise with the selective etiological theory of function. But it would probably make it a little bit more complicated.

References

Allen, C., M. Bekoff, and G. Lauder (eds.). 1998. *Nature's purposes. Analyses of function and design in biology*. Cambridge: The MIT Press.

Ariew, A., R. Cummins, and M. Perlman (eds.). 2002. *Functions. New essays in the philosophy of psychology and biology*. Oxford: Oxford University Press.

Ayala, F. 1968. Biology as an autonomous science. *American Scientist* 56(3): 207–221.

Bigelow, J., and R. Pargetter. 1987. Functions. *The Journal of Philosophy* 84: 181–197.

Blandin, P. 2007. L'écosystème existe-t-il ? Le tout et la partie en écologie. In *Le tout et les parties dans les systèmes naturels*, ed. T. Martin, 21–46. Paris: Vuibert.

Block, N. (ed.). 1980. *Readings in philosophy of psychology*, vol. I. Cambridge: Harvard University Press.

Bouchard, F. 2008. Causal processes, fitness and the differential persistence of lineages. *Philosophy of Science* 7: 560–570.

Buller, D.J. (ed.). 1999. *Function, selection and design*. Albany: State University of New York Press.

Cummins, R. 1975. Functional analysis. *The Journal of Philosophy* 72: 741–765.

Gayon, J., and A. de Ricqlès. 2010. *Les fonctions: des organismes aux artéfacts*. Paris: Presses Universitaires de France.
Godfrey Smith, P. 1994. A modern history theory of functions. *Noûs* 28: 344–362.
Lorne, M.C. 2004. Explications fonctionnelles et normativité: analyse de la théorie du rôle causal et des théories étiologiques de la fonction. Philosophy Ph.D. thesis. Paris: EHESS.
McLaughlin, P. 2001. *What functions explain. Functional explanation and self-reproducing systems*. Cambridge: Cambridge University Press.
Neander, K. 1983. Abnormal psychobiology. A thesis on the 'anti-psychiatry debate' and the relationship between psychology and biology. Ph.D. thesis, La Trobe University, Philosophy Department.
Neander, K. 1991. The teleological notion of function. *Australasian Journal of Philosophy* 69: 454–468.
Pacherie, É. (éd.). 1995. Fonctionnalismes. *Intellectica*, n° spécial, 21, 1995/2.
Van Valen, L. 1991. Biotal évolution: a Manifesto. *Evolutionary Theory* 10: 1–13.
Vermaas, P., and W. Houkes. 2003. Ascribing functions to technical artefacts: A challenge to etiological accounts of functions. *The British Journal for the Philosophy of Science* 54: 261–289.
Wimsatt, W. 2002. Functional organization, analogy, and inference. In *Functions—New essays in the philosophy of psychology and biology*, ed. A. Ariew, R. Cummins, and M. Perelman, 173–221. Oxford: Oxford University Press.
Wright, L. 1973. Functions. *Philosophical Review* 92: 139–168.
Wright, L. 1976. *Teleological explanations: An etiological analysis of goals and functions*. Berkeley: University of California Press.

Part II
Biological Functions and Functional Explanations: Genes, Cells, Organisms and Ecosystems – Functional Pluralism for Biologists?

How Ecosystem Evolution Strengthens the Case for Functional Pluralism

Frédéric Bouchard

Abstract Evolutionary explanations appear to necessitate etiological theories of function. As Amundson and Lauder have shown (Amundson R, Lauder GV. Function without purpose: the uses of causal role function in evolutionary biology. *Biol Philos* 9:443–70, 1994, reprinted in Allen et al. Nature's purposes analyses of function and design in biology. The MIT Press, Cambridge, MA, 1998), current biological practice is in fact more pluralistic in its choice of functional explanations, using etiological functions as well as ahistorical causal functions. Here, I will examine how some functional descriptions in ecology and how they are imported into evolutionary explanations strengthen the case for the use of ahistorical functional theories in biology in general but in ecology and evolutionary biology in particular. I will focus on the case of ecosystem evolution where I will argue that fitness is better understood as differential persistence. We shall see that this type of evolutionary phenomenon demands nonhistorical functional explanations. This will be described as a potential vindication for forward-looking functional theories, otherwise known as propensity account of functions. In a more general way, I will show how this vindicates pluralistic account of functional explanations in biology.

1 Introduction

In this chapter, we will examine how some functional descriptions in ecology and how they are imported into evolutionary explanations strengthen the case for the use of ahistorical functional theories in biology in general but in ecology and evolutionary biology in particular. I will focus on the case of ecosystem selection and evolution

F. Bouchard (✉)
Department of Philosophy, Université de Montréal, Montréal, QC, Canada
e-mail: f.bouchard@umontreal.ca

P. Huneman (ed.), *Functions: selection and mechanisms*, Synthese Library, Studies in Epistemology, Logic, Methodology, and Philosophy of Science 363,
DOI 10.1007/978-94-007-5304-4_5,

and how it demands nonhistorical functional explanations. This will be described as a potential vindication for forward-looking functional theories, otherwise known as propensity account of functions. In a more general way, I will argue that this vindicates pluralistic account of functional explanations in biology.

The discussion about functions in biology has focused mainly on evolutionary biology for a few distinct reasons. Let us briefly examine two of these reasons.

First, philosophers of mind were looking for a way to offer some sort of teleological or quasi-teleological grounding for functional ascriptions in a way that constrained the types of structures that could instantiate the functional systems. Teleofunctionalism in philosophy of mind seemed able to evacuate problems that functionalism had with, for example, inverted qualia, or malfunctioning traits in general. This teleofunctionalism was cashed out in evolutionary terms.

Another reason for the focus of functional arguments on evolutionary biology is that, and this is more a sociological point than a philosophical one, most philosophers of biology for the last 40 years have focused their inquiry evolutionary theory, in part because it appears prima facie to be the best candidate for a unifying theory of biological explanations, something that developmental biology or cell biology cannot hope to achieve.

In this context, it is not surprising that theories of function grounded on history have been favoured: evolutionary history appears to get philosophy of mind out of theoretical binds while warranting philosophy of biology's focus on evolutionary biology instead of focusing on other biological disciplines.

What is not always recognized however is that evolutionary history will not always vindicate historical functional theories. Even though evolutionary biology has to look at selection history which seems to warrant a Wright-like function theory (Wright 1973, 1976), some evolutionary explanations necessitate nonhistorical functional ascriptions. For most readers of this book, this is probably not a novel point although it is still somewhat controversial. Amundson and Lauder in their oft-quoted 1994 paper (reprinted in Allen et al. 1998) argue for functional pluralism, that is, we should entertain both historical and nonhistorical functional theories since they are both necessary for biological discovery; Amundson and Lauder use examples from physiology and other biological fields that cannot be said to use historical functional concepts and show that they are necessary for evolutionary explanations. More recently, Griffiths (2006) has offered a similar argument using developmental biology (although he uses this example not to defend a pluralist view but something more akin to a monist nonhistorical functional theory). After briefly discussing this pluralist line of argument, I will use examples stemming from ecology to show that nonhistorical functions are necessary for biological explanations.

But I will add a twist; I will show how these nonhistorical functions are necessary not only for ecology but for evolutionary biology and play an even larger role than what was described by Amundson and Lauder. This result is somewhat ironic since it would show that some aspects of evolutionary explanations do not depend on past history… Propensity accounts (or dispositional, or forward-looking accounts) of functions are truly necessary, which reduces (but does not eliminate) the relative importance of historical functions in evolutionary explanations. As we shall see, this point has been made before, but the support provided here is novel and more importantly shows the scientific urgency of thinking about these questions.

2 Diversity Rules

Wright (1973, 1976) famously described an etiological thesis where the function must be understood as an explanation of the persistence of an entity through time. As many have pointed out (e.g. Boorse 1976; Godfrey-Smith 1998; Millikan 1989), notice that this description does not entail that historicity only concerns biological entities although evolutionary theory puts this historicity in a plausible natural context. Non-biological entities also exist for a period of time; anything that contributes to the object's continuing existence is to be considered functionally. Millikan's account of proper functions (Millikan 1989, 1993) may be a good example of Wright functions but one that wishes to be geared towards biological entities. Millikan, herself, rejects the idea that she is merely exposing a refined Wright function.[1]

The difference between the two theories would be concerning the concept of origin. According to Millikan, Wright speaks of etiology without talking about specific origins of the entities and their functions. Whether this is enough to distinguish Millikan's thesis from Wright's is arguable. I am inclined to say that Millikan's proper functions are but a special case of Wright functions: this special instantiation could be seen as purely biological. This point will not be examined further here. Whatever the degree of similarity or affinity between Wright's and Millikan's account, the fact remains that both rely on a notion of past selected effects and therefore on processes that have unfolded in the past, that is, historically. Neander and Godfrey-Smith (among others) have added further precisions to this account but the details will not concern us here.

The other functionalist camp rejects this historicity. Cummins (1975) argues that it is the *now* that science is interested in, and as such, it would be misleading to understand functions exclusively relative to their origin. Evolutionary theory is not needed to identify biological function. This is not surprising per se since Cummins himself is interested only in the concept of function as it is used in psychology, but others have used his concept in biology. In Cummins' view, functional explanations reflect the contribution of a capacity to the overall capacity of the system. The understanding of such capacity is ahistorical because actual capacities do not necessarily reflect the original goal or the purpose of the system. This means that we should only examine how the system is working at moment t1 and try to figure out how the different parts of the system work together at t1.

[1] See a comment to that effect in note 5 of Millikan (1989).

According to this functional theory dichotomy, one may prima facie believe that evolutionary biologists would focus on historical Wright functions (because of the focus on evolutionary history) and other biologists would focus on ahistorical Cummins functions.

This dichotomy (at least in the case of evolutionary biology) may be overly simplistic. Amundson and Lauder show that both functional theories underpin evolutionary explanations, that is, it is false that evolutionary biology only concerns itself with historical functions.

The bias of many evolutionary biologists (or rather, philosophers interpreting biological theory) is to see functions in evolutionary explanations exclusively as Wright functions, what Amundson and Lauder call selected effects (SE) functions. Amundson and Lauder defend the idea that evolutionary biology is also concerned with Cummins functions that they call causal role (CR). By describing subfields of evolutionary biology that do not put any relevance in the SE thesis, Amundson and Lauder show that CR is necessary for evolutionary biology. This is significant because, if Amundson and Lauder can show that some evolutionary biology research cannot be served exclusively by SE, and actually sometimes doesn't require it, and that, inversely, some other evolutionary biologists cannot do without SE, a pluralistic functional account will be necessary to account for functional explanations in evolutionary biology: different functionalisms will be needed in different subfields. It appears that this conclusion is intended both as a descriptive claim (i.e. evolutionary biology *is* actually pluralistic) and a normative claim (i.e. evolutionary biology *should be* pluralistic with regard to functional explanations).

As an example of a CR proponent, Amundson and Lauder use the case of functional anatomists who look at bone structures and organisms ahistorically – they consider all the possible capacities of a structure in an engineering-like way. If we accept the relevance of their work and more importantly the necessity of this work, for example, in the trait identification in palaeontology, we must accept that SE functions will not be sufficient in evolutionary biology. Some of the critiques of the CR view have questioned the antecedent in the previous conditional by questioning the relevance of functional anatomists. CR functions are painted as playing with trivial hypothetical descriptions. Amundson and Lauder show that this characterisation is unfair: functional anatomists, while considering possible capacities, are examining possible capacities of *actual* systems. Their explanations are not the trivial description of science-fiction cases as their opponents would make them out to be. The example I will describe later will hopefully be another argument in favour of nonhistorical functional analyses.

One must stress the point that Amundson and Lauder are not rejecting a SE view of function. Rather, they are arguing that an exclusive SE view gives an impoverished view of the field of evolutionary biology.

Conversely, they argue, a purely CR view of function cannot do the whole job. That is the reason a pluralistic account of function is needed, one where SE is useful in certain cases and CR is useful in others. Amundson and Lauder argue that reducing one to the other doesn't give a true characterisation of evolutionary biology as a whole.

3 Looking Ahead

As it has often been pointed out, Wright wasn't concerned with evolution per se. In fact, his descriptions of functions were devoid of any biological criteria, even though they would be compatible with a biological framework. The non-biological framework Wright described showed certain weaknesses. Boorse showed[2] that the way Wright functions are construed, one could ascribe trivial functions to systems that could be described as having a 'purpose', but that would be described as such only because of circumstantial evidence.

Take Wright's definition of function:

> The function of X is Z *means* that (a) X is there because it does Z. (b) Z is a consequence (or result) of X's being there. (Wright 1973, p.161)

Boorse, Godfrey-Smith and Millikan among others note that 'the problem here is with the broad range of "X" and "Z" [which are the variables in Wright Functions]' (Godfrey-Smith in Allen et al. 1998, p. 455). Without any biological criteria (or in fact any other type of criteria), there is no way to determine what are the relevant entities that need to be explained functionally and what functional explanations are not trivial. Millikan, for example, wishes to use biology insofar as it constrains the domain of application of functional inquiry in a meaningful way. By doing so, we eliminate a priori many trivial cases of hypothetical teleology.

As previously noted, SE functions (and the explanation they provide) are not rejected by Amundson and Lauder. Rather, they argue for the importance of physiology and functional anatomy and the nonhistorical functional explanations they provide and their significant role in evolutionary explanations. For SE functions, one needs a past history of selection to identify the process and its 'real' function. As Amundson and Lauder point out, a purely engineering view is sometimes necessary when history is not available. But is that merely an epistemic point? The fact that our study of the fossil record, because of its relative poor quality, leads to nonhistorical description may say more about our access to evidence than about functional explanations per se.

In other words, a genuine worry is that functional pluralism might only be a temporary state of affairs: given more information about the living world, we could eliminate the instrumental use of nonhistorical causal functions and revert to historical functions simpliciter. Basically CR functions could be seen as instrumentally necessary for now, but ultimately disposable in favour of the 'real' SE functions.

I will now offer some hope that this worry is overstated: at least in some biological cases, the use of some sort of nonhistorical functions is not merely instrumental and does not merely depend on our epistemic constraints.

I wish now to examine an evolutionary case where there is no past history.

The problem of past versus future history is the core of the problem here.

---

[2] Boorse (1976), or see Griffiths in Allen et al. (1998), p. 445 for a detailed summary of the argument and a thoughtful discussion of this issue.

As Bigelow and Pargetter (1987 reprinted in Allen et al. 1998) point out, SE functions are purely backward-looking descriptions. A given trait has a specific function if that function contributed *in the past* to the persistence of that trait. But as they point out, some sort of forward-looking accounts play a large role in conventional accounts of fitness. At their core, propensity accounts of *fitness* are causal accounts. The probability of an organism to have a certain number of offspring is grounded on the physical, biological and behavioural features of the organism and how it interacts, causally interacts that is, with its environment. But propensity accounts are interested in probable offspring contribution, not actual offspring contribution. In the same way that propensities allow fitness to avoid the tautology problem, Biggelow and Pargeter argue that a propensity account of functions gives the explanatory force of functional explanations.

And they come up with this suggestive conclusion:

> The etiological theory describes a character now as serving a function when it did confer propensities that improved the chances of survival. We suggest that it is appropriate, in such a case, to say that the character *has been serving that function all along*. Even before it had contributed (in an appropriate way) to survival, it had conferred a survival-enhancing propensity on the creature. And to confer such a propensity, we suggest, is what constitutes a function. Something has a (biological) function just when it confers a survival-enhancing propensity on a creature that possesses it. (Bigelow and Pargetter in Allen et al. 1998, p. 252)

Similar accounts have been given by Wimsatt (1972) for instance. The nice thing about propensities is that for better or for worse, one does not need past history. One could have a propensity even if a system and its functions appeared ex nihilo. This is not the case for SE functions and this will be crucial for the rest of my argument.

I will now show how ecosystem evolution can be understood and how, because of the abiotal part of ecosystems, one needs some sort of nonhistorical account of functions and of fitness. As it will become clear in the following pages, we will be relying on highly unorthodox ways of understanding evolution by natural selection. Yet, the hope is that payoff of adopting them outweighs the cost of changing our evolutionary framework.

Leo Buss's description of somatic selection (Buss 1983) is an inspiration for this part of the argument: Weismannism describes how only changes in the germ line can be passed on to the next generations. But as Buss points out convincingly, the evolution of protists, fungi and some plants which are in large part the result of selection on somatic changes cannot be accommodated by Weismannism. Buss uses this idea to justify a hierarchical view of selection broader than the usual modern synthesis view. Many of the examples given by Buss literally do not reproduce. Buss is correct in explaining how, in the cases he presents, evolution can happen via selection on sub-organismal variation. As we will see for some cases of evolution, the notion of component or part is more relevant than the notion of offspring. This insight has found some support in more orthodox understanding of evolutionary theory.

In his exhaustive survey of natural selection experiments, John Endler (1986) pointed out that many studies in evolutionary biology focus exclusively on intragenerational success and phenotypic selection. Although obviously fecundity and fertility are keystones of evolutionary explanations, survival and the means by which organisms survive are a necessary aspect of the story.

Elsewhere (Bouchard 2004, 2007, 2008, 2011) I argue in details that *differential persistence* should replace *differential reproductive success* for a unified understanding of fitness. I can't give the whole argument here but the broad motivation is straightforward: what is necessary is a broader understanding of evolution to cover the evolution of strange entities like corals, huge integrated clones and, the example I will examine here, ecosystems.

This insight is inspired in part by Van Valen:

> It is just as good, and maybe better, for a massive coral or a tree to stay alive, occupying the same good site, as it is for it to reproduce into an uncertain world.
>
> (…)
>
> Persistence is an important component of fitness and is ultimately related to the spatiotemporal heterogeneity of the total environment. (Van Valen 1989, p. 5)

For many biological systems, differential success does not perfectly match differential reproductive success. This is a controversial claim, especially since allelic frequencies are the current key metric of adaptive success in our evolutionary explanations. Yet the problems are well known: for many plants, for example, it has always been difficult to distinguish asexual reproduction that can count as differential reproductive success, from vegetative growth that concerns development more than evolution. Philosophers have assessed this difficulty by arguing that reproductive success while central may not be exhaustive to account for evolutionary success. Ariew and Lewontin (2004) have highlighted the problem of asexual reproduction for a reproductive-based account of fitness while Sober (2001) has described the dual understanding of fitness, the first usually focusing on reproductive success, while the other facet focuses on survival. In my previous work, I develop this last aspect to encompass all others. Fitness is usually understood as a composite of survival and reproduction, yet, in most models, survival is only included as instrumentally necessary to get the organism to the reproductive phase. I turn this relationship on its head to argue that reproduction is a means to increase the lineage's persistence (equivalent of survival). This idea is inspired by similar moves stemming from ecology.

The focus on persistence has been around for a long time in ecology (often under the guise of stability). Persistence was not seen by most ecologists as an evolutionary property. This is not surprising given that ecosystems do not have their own genetic systems (and therefore heritability at the ecosystem level is prima facie a non-starter). But, once one identifies ecosystem-level property (e.g. stability, complexity, species-richness), it is but a small leap to hypothesize that this property is the result of selection-like forces. Ecosystems obviously do not reproduce but they do persist, some better than others, giving us the building blocks of differential success. Many advocates of the idea that whole ecosystems could evolve quickly realize that persistence, not reproduction, will be the key to understand ecosystem evolution.

Theoretically, the idea of ecosystem evolution is interesting but the problem has always been to identify real cases of ecosystem evolution. Ecosystem evolution had until very recently not been identified as a genuine evolutionary process (although many believed it was at least a theoretical possibility). It was believed to be epiphenomenal (Hoffman 1979) or at least very unlikely (Hull 1980). But within ecology, the judgement has not been so pessimistic. A few texts stand out as evolutionary descriptions of ecosystem creation, maintenance and transformation. Ott (1981) in his assessment of marine ecosystem writes that 'Although the basic features of evolution can be found in ecosystem development, the mechanism is quite different from Darwinian evolution. Ecosystem fitness is not determined by differential reproduction but rather by differential persistence (survival)' (Ott 1981, p. 144). Dunbar (1960), also focusing on marine ecosystems, arrives at a similar conclusion. 'As to the mechanisms by which selection might take effect at this [ecosystem] level, they are of the ordinary Darwinian sort except that the criterion for selection is survival of the system rather than of the individual or even the species'(Dunbar 1960, p. 134). Cropp and Gabric (2002) focus on the evolution of resilience as an ecosystem-level adaptation. Darnell (1970) goes further by placing ecosystem evolution at the heart of all evolutionary process. Many other ecologists have entertained the idea that ecosystems can evolve by natural selection, but this research programme is fraught with obstacles.

Part of the operational difficulty in testing the ecosystem evolution hypothesis is a problem of physical scale. How can one go about 'measuring' the evolutionary fate of whole ecosystems? Ecosystems are usually construed as relatively large, and it is very difficult to account for all the species constituting it and the interactions between them. But when one realizes that ecosystem or communities to not have to be 'large' relative to human scale, testing evolutionary hypotheses becomes much more manageable.

In artificial selection experiments (Swenson et al. 2000a, b), a good case for artificial ecosystem selection is provided. I will refer to the experiments as David Sloan Wilson's experiments since he was, as far as I can understand it, the principal investigator in all three studies. Wilson and others describe three experiments where artificial selection is used to shape the phenotype of whole ecosystems. In all cases, they use mud samples and try to select for a certain phenotype.

Let me briefly describe one of their experiments.

They take 2 ml of sediment (full of dirt, bacteria, etc.) and 28 ml of water from a pond and fill 72 test tubes, which are then incubated. Each tube is then measured for pH level, which was the arbitrary trait they decided to select on, but a good trait to measure phenotypic change in ecosystems since the pH level is a feature of the physical substrate, the dirt, and the water, as well as a phenotype of the microorganisms living in the dirt. They then take the six test tubes with the highest pH. From each of these six test tubes, they take 5 ml of mud and add 25 ml of autoclaved pond mixture. And repeat. They observed an increase in pH level in the 'winning test tubes'. As strange as it seems, the mud samples produced the phenotype that enabled them to 'survive' in this artificial selective environment. And more importantly, the phenotypes were stable enough so that the increase in pH level actually was retained across 'generations' and amplified across time.

By showing how small malleable ecosystems could be artificially selected to 'get' a particular trait, they show that at least in theory, we could observe the same thing in nature. Goodnight (2000) and Penn (2003) examined the heritability involved in these experiments (focusing on the community aspect more than the ecosystemic nature of the system), while Williams and Lenton (2007) reprise this idea to assess evolutionary optimization in ecosystems.

Many ecologists have focused on energy transfers/control or on entropy in general in ecosystems and how selection can act on ecosystems to maximize this control. This is implicit in Fath et al. (2004), explicit in Van Valen (1991) and offers for Loreau (2010), in his rich volume on the desirable dialogue between community ecology and ecosystem ecology, the best hope of unifying ecology and evolution (see also Felsenstein 1978; Fussmann et al. 2007 and Loreau 2009). In Bouchard (2004), I argue why energy control while offering a common-currency control for fitness has its own disadvantages. I focus instead on differential persistence of the system, but the idea remains that ecosystems can evolve.

To make sense of ecosystem evolution, defining fitness in terms of offspring numbers will only take us so far. There is internal competition between microorganisms in the mud sample, but they argue that the causal explanation at the ecosystem level however remains: microsystems with higher pH persisted better than microsystems with lower pH. The pH level is a trait of the ecosystem and a trait of the whole system is selected for.

The only way for the 'mud' to persist is if it changes its pH (the teleological connotation is merely a manner of speaking), and it does so without reproducing. But its phenotype changes thanks to environmental pressures, and this change persists and increases over time.

Again I am not claiming that reproduction is not involved at all here, but I am claiming that it is not the salient feature to explain the transformation of the phenotype of the ecosystem as a whole. Think of it this way. Let's say that a higher pH lead to slower erosion. The patches of mud with a higher pH would persist, whereas the ones with lower pH would erode. There is natural selection here. But is there evolution? If the patch only gets smaller and smaller, there is just natural selection.[3] If the patch eventually stabilizes, and moreover may grow thanks in part to reproductive success of some of its microorganisms but also possibly to the chemical reactions of the physical substrates AND if the pH increases (leading to less erosion), then it seems we have evolution by

[3] This is not surprising in itself since, as Van Valen points out, even non-biological structures may be subject to natural selection 'When granite weathers, the feldspars and micas become clays but nothing much happens to the quartz grains. They are most resistant and get transported down streams or along shores. Thus most beaches are the result of differentially eroded granite. This is an example of natural selection in the nonliving world. Quartz grains survive longer than feldspar grains, and there is a progressive increase in the average resistance to weathering, of the set of grains that have still survived. This action of natural selection is even creative, as we see by the formation of a beach' (Van Valen 1989, p. 2).

natural selection even though offspring contribution might not be the best way to describe the evolutionary change. But intuitively, we have a way to define the fitness of that patch. It 'offered' a better solution to a design-problem! It can still be a propensity (a propensity to have a higher pH in this case), but it isn't defined in offspring contribution since the patch may expand (or minimally persist) without really reproducing. To understand the fitness of the ecosystem, one will have to understand how components of that ecosystem contribute to the capacity to persist.

Thoday in 1953 suggested that to be fitter is to have a higher propensity to leave at least one offspring in 10^8 years offering an understanding of fitness grounded in long-term persistence. But why should we talk about offspring at all? If we wish to examine two ecosystems, couldn't we compare their relative fitness in terms of their capacity to still be there in x number of years? Couldn't we say that *if* this propensity (which will fluctuate over time) is the result of environmental pressures, then what we have is evolution by natural selection? Ecologists have been suggesting concepts like differential persistence for ecosystems for many years. My suggestion is to extend this to other evolutionary phenomena.

Not surprisingly, this comes very close to the definition of function offered by Bigelow and Pargetter (1987 and reprinted in Allen et al. 1998). They focus on individual survival, but persistence is the more general feature of interest here.[4]

If one wishes to understand the evolution of ecosystems, one will have to explain the role of various components of those ecosystems. The biotal component of these components may be explainable via SE functions – after all they are the result of descent with modification of other species. It's not obvious however that community evolution (i.e. interaction between different species) will always have such past histories. But more crucially, ecosystems are not just biotal (i.e. living) material. As the mud case hints, ecosystems are also geological, chemical and physical in nature. This means that significant components of these evolving entities cannot have SE functional role even though they may play a crucial role that will explain the capacity of that given ecosystem to persist longer than other similar ecosystems. With ecosystems, we have entities that may be evolving, do so *sans* differential reproductive success and where differential persistence is the measure of evolutionary success. More importantly, ecosystems are entities whose components do not always have SE functions in the strictest sense. However, these components are a necessary part of the explanation of ecosystems' increased persistence. Therefore, SE functions are not sufficient to understand the functions of subsystems in evolving ecosystems.

In Swenson et al.'s example, we have a feature of a system (here the increased pH level of an ecosystem) that is the result of changes in selection pressures. Such feature does not have a past history, although it may have a 'bright' future … so to ascribe functional explanations, one needs some type of engineering analysis to make sense of the functioning of the system.

[4] As a side remark, thinking in terms of persistence instead of survival might help them extend their framework to artefacts, which is something they hope to achieve….

4 Conclusion

Is this merely another item on the list of items that cannot be accounted for by SE functions alone? Well yes and no.... If one retraces some of the history of the vindication of CR functions, one could say that Amundson and Lauder started by showing that physiology and functional anatomy exclusively use CR functions. Then Griffiths argued that developmental biology was another CR discipline. One of my goals is to add ecology as another functional orphan (relative to SE accounts)....

But there is more to the story than this. The claim here is more subtle. Following ecologists' theoretical work and recent empirical work, I am claiming that ecosystems evolve, but these ecosystems will not be part of lineages (as they are usually construed).... One could argue that some ecosystems may have been evolving for a long time, and the succession of states they have gone through will in some sense constitute some sort of lineage.

But more controversially, I would argue that new ecosystems 'appear' all the time and will start evolving. A landslide creates new ecosystems that will respond to selective pressures and could possibly evolve. A hurricane will redraw marshes and put species in new relationships. When trying to understand how these ecosystems evolve, one will not have access to past history and selected effect to understand the various functions of components of ecosystems. And this is not an epistemic blind spot as the case of the fossil record. It is the result of the coming into being of new entities. Ecosystems appear and disappear in much more transient fashion than other biological systems do.

The problem of novelty has always been a genuine worry for evolutionary biology. One can use evolution by natural selection to explain the maintenance and the transformation of a given trait, but it's not obvious how completely novel traits can appear (and they must at some point). This problem inspired many to argue for an increased look at developmental biology and its fusion with evolutionary biology in evo-devo. This is in part what motivates Griffiths to entertain nonhistorical functions. With ecosystem evolution, we seem to get the novelty problem in spades: new ecosystems and new components of ecosystems without any selection history appear all the time. Since those ecosystems may be evolving, it means we need, *at least for these cases*, nonhistorical functional explanations in evolutionary explanations: propensity accounts of function like the one suggested by Bigelow and Pargetter might be a good candidate. Ironically, what Bigelow identify has a possible pitfall of their account is exactly the type of opportunity I wish to explore. After describing the advantages of their account, they identify some 'less comfortable results'.

> Suppose a structure exists already and serves no purpose at all, Suppose then that the environment changes, and, as a result, the structure confers a propensity that is conducive to survival. Our theory tells us that we should say that the structure now has a function (Bigelow and Pargetter in Allen et al. 1998, p. 246).

Of course, this whole discussion is moot if ecosystems cannot in fact evolve. But as I have pointed out, promising empirical results indicate that they can. Ecosystems display adaptive change as a response to the selective environments, and these changes accumulate and are fine-tuned over time in order to increase the system's capacity to survive. However, these systems' evolution is not adequately captured by a concept of evolutionary fitness that is defined solely in terms of differential reproductive success, and a fortiori it will be difficult to make sense of intergenerational change. More importantly new ecosystems come into being all the time. To make a truly bad analogy (and a worse jeu de mot in this context), we have the equivalent of philosophy of mind's Swampman (Davidson 2001). If this is the case, one will need some sort of nonhistorical functional description to understand how they work and how they evolve.

The claim here is not that nonhistorical functions are sufficient for evolutionary explanations, but rather that ecosystem evolution vindicates some sort of functional pluralism in biology: we can use nonhistorical functional explanation as the only foundation of some evolutionary explanations when there is no available history (again not merely an epistemic point like in some of Amundson and Lauder's examples, but a metaphysical point: there exists no history).

As some of you may know, Leigh Van Valen often ends his talks with a song. It is only appropriate to end from a line from a song that he recommended to me when we were discussing these issues.

The Hippopotamus Song by Flanders and Swann

Mud! Mud! Glorious mud!
Nothing quite like it for cooling the blood.
So, follow me, follow, down to the hollow,
And there let us wallow in glorious mud.

Maybe mud can help us better understand fitness and functions as well.

References

Allen, C., M. Bekoff, and G.V. Lauder. 1998. *Nature's purposes analyses of function and design in biology*. Cambridge, MA: The MIT Press.

Amundson, R., and G.V. Lauder. 1994. Function without purpose: The uses of causal role function in evolutionary biology. *Biology and Philosophy* 9: 443–470.

Ariew, A., and R.C. Lewontin. 2004. The confusion of fitness. The British Journal for the Philosophy of Science 55: 365–370.

Bigelow, J., and R. Pargetter. 1987. Functions. *Journal of Philosophy* 84: 181–196.

Boorse, C. 1976. Wright on functions. *Philosophical Review* LXXXV(1): 70–86.

Bouchard, F. 2004. Evolution, fitness and the struggle for persistence. Ph.D. thesis, Duke University.

Bouchard, F. 2007. Ideas that stand the [evolutionary] test of time. In *Interdisciplines: Adaptation and representation*. Paris: CNRS. http://interdisciplines.org/adaptation/papers/12

Bouchard, F. 2008. Causal processes, fitness and the differential persistence of lineages. *Philosophy of Science* 75: 560–570.

Bouchard, F. 2011. Darwinism without populations: A more inclusive understanding of the "Survival of the Fittest". *Studies in History and Philosophy of Science Part C: Studies in History and Philosophy of Biological and Biomedical Sciences* 25(4): 623–641.

Buss, L.W. 1983. Evolution, development, and the units of selection. *Proceedings of the National Academy of Sciences of the United States of America* 80(5, [Part 1: Biological Sciences]): 1387–1391.

Cropp, R., and Albert Gabric. 2002. Ecosystem adaptation: Do ecosystems maximize resilience? *Ecology* 83(7): 2019–2026.
Cummins, R. 1975. Functional analysis. *Journal of Philosophy* 72: 741–765.
Darnell, R.M. 1970. Evolution and the ecosystem. *American Zoologist* 10(1): 9–15.
Davidson, D. 2001. *Subjective, intersubjective, objective (Philosophical essays of Donald Davidson)*. Oxford: Oxford University Press.
Dunbar, M.J. 1960. The evolution of stability in marine environments natural selection at the level of the ecosystem. *The American Naturalist* 94(875): 129–136.
Endler, J.A. 1986. *Natural selection in the wild*. Princeton: Princeton University Press.
Fath, B.D., et al. 2004. Ecosystem growth and development. *Biosystems* 77: 213–228.
Felsenstein, J. 1978. Macroevolution in a model ecosystem. *The American Naturalist* 112(983): 177–195.
Fussmann, G.F., M. Loreau, and P.A. Abrams. 2007. Eco-evolutionary dynamics of communities and ecosystems. *Functional Ecology* 21(3): 465–477.
Godfrey-Smith, P. 1998. *Complexity and the function of mind in nature*. Cambridge/New York: Cambridge University Press.
Goodnight, C.J. 2000. Heritability at the ecosystem level. *PNAS* 97(17): 9365–9366.
Griffiths, P.E. 2006. Function, homology, and character individuation. *Philosophy of Science* 73(1): 1–25.
Hoffman, A. 1979. Community paleoecology as an epiphenomenal science. *Paelobiology* 5(4): 357–379.
Hull, D.L. 1980. Individuality and selection. *Annual Review of Ecology and Systematics* 11: 311–332.
Loreau, M. 2009. Linking biodiversity and ecosystems: Towards a unifying ecological theory. *Philosophical Transactions of the Royal Society B: Biological Sciences* 365(1537): 49–60.
Loreau, M. 2010. *From populations to ecosystems: Theoretical foundations for a new ecological synthesis (MPB-46)*. Princeton: Princeton University Press.
Millikan, R.G. 1989. In defense of proper functions. *Philosophy of Science* 56(2): 288–302.
Millikan, R.G. 1993. *White Queen psychology and other essays for Alice*. Cambridge, MA: MIT Press.
Ott, J.A. 1981. Adaptive strategies at the ecosystem level: Examples form two benthic marine systems. *Marine Ecology* 2: 113–158.
Penn, A. 2003. Modelling artificial ecosystem selection: A preliminary investigation. In *Advances in Artificial Life*, Lecture Notes in Computer Science. 2801: 659–666. Springer Berlin/ Heidelberg.
Sober, E. 2001. The two faces of fitness. In *Thinking about evolution: Historical, philosophical, and political perspectives*, ed. Rama S. Singh, Costas B. Krimbas, Diane B. Paul, and Beatty John, 309–321. New York: Cambridge University Press (xvii, 606 p).
Swenson, W., J. Arendt, and D.S. Wilson. 2000a. Artificial selection of microbial ecosystems for 3-chloroaniline biodegradation. *Environmental Microbiology* 2(5): 564–571.
Swenson, W., D.S. Wilson, and R. Elias. 2000b. Artificial ecosystem selection. *Proceedings of the National Academy of Sciences of the USA* 97(16): 9110–9114.
Thoday, J.M. 1953. *Components of fitness' symposia of the society for experimental biology*, 96–113. Cambridge: Cambridge University Press.
Van Valen, L.M. 1989. Three paradigms of evolution. *Evolutionary Theory* 9: 1–17.
Van Valen, L.M. 1991. Biotal evolution: A Manifesto. *Evolutionary Theory* 10: 1–13.
Williams, H., and T. Lenton. 2007. Artificial selection of simulated microbial ecosystems. *Proceedings of the National Academy of Sciences* 104: 8918–8923.
Wimsatt, W. 1972. Teleology and the logical structure of function statements. *Studies in the History and Philosophy of Science* 3: 1–80.
Wright, L. 1973. Functions. *Philosophical Review* 82(2): 139–168.
Wright, L. 1976. *Teleological explanations: An etiological analysis of goals and functions*. Berkeley: University of California Press.

A General Case for Functional Pluralism

Robert N. Brandon

Abstract Using examples from functional morphology and evolution, Amundson and Lauder (Biol Philos 9: 443–469, 1994) argued for functional pluralism in biology. More specifically, they argued that both causal role (CR) analyses of function and selected effects (SE) analyses played necessary parts in evolutionary biology, broadly construed, and that neither sort of analysis was reducible to the other. Rather than thinking of these two accounts of function as rivals, they argued that they were instead complimentary. Frédéric Bouchard (Chap. 5, this volume) attempts to make that case stronger using an interesting example—the evolution of ecosystems. This case is interesting in that it involves the sudden appearance of things with functions, which also evolve, but which do not, at least initially, have a selected effect etiology. I am in complete agreement with the above-mentioned positions. Here, I take a different tack in arguing for functional pluralism. I abstract away not only from the details of biological practice but even from the details of the CR and SE accounts to argue for a more general pluralism of historical and ahistorical concepts.

Using examples from functional morphology and evolution, Amundson and Lauder (1994) argued for functional pluralism in biology. More specifically, they argued that both causal role (CR) analyses of function and selected effects (SE) analyses played necessary parts in evolutionary biology, broadly construed, and that neither sort of analysis was reducible to the other. The SE account of function is explicitly historical. According to it, an item has a particular function if and only if it owes its current form and frequency (in a population, a species, or in a clade)

My thanks to Karen Neander for helpful comments on an earlier draft of this chapter.

R.N. Brandon (✉)
Departments of Biology and Philosophy, Duke University,
Durham, NC 27708, USA
e-mail: rbrandon@duke.edu

P. Huneman (ed.), *Functions: selection and mechanisms*, Synthese Library,
Studies in Epistemology, Logic, Methodology, and Philosophy of Science 363,
DOI 10.1007/978-94-007-5304-4_6,

to natural selection acting on it in virtue of its having the effect identified as its function.[1] In contrast, the CR account of function is ahistorical. A CR function is an effect of the item in question, say a wing, which helps explain how that item contributes to some capacity, say flight, of some larger containing system, say a bird.[2] Rather than thinking of these two accounts of function as rivals, Amundson and Lauder (1994) argued that they were instead complementary. Frédéric Bouchard (Chap. 5, this volume) attempts to make that case stronger using an interesting example—the evolution of ecosystems. This case is interesting in that it involves the sudden appearance of things with functions, which also evolve, but which do not, at least initially, have a selected effect etiology. I am in complete agreement with the above-mentioned positions. Here, I take a different tack in arguing for functional pluralism. I abstract away not only from the details of biological practice but even from the details of the CR and SE accounts to argue for a more general pluralism of historical and ahistorical concepts. To do this, I first turn away from biology and toward the geology of mountains.

1 Mountain Geology

Mountains are formed by two fundamentally different geological processes. The first is the relative movement of tectonic plates. Two dramatic examples of that are the Alps, which were formed when the African plate pushed northward colliding into the European plate (in the late Cretaceous period), and the Himalayan and Karakorum ranges that form an arc above the Indian subcontinent. They too are the result of continental collision, in this case the northward movement of the Indian plate into the Eurasian plate (at about the same time as the formation of the Alps). The Alps and the Himalayas are among the youngest mountain ranges on Earth, their morphology being the result of the folding of Earth layers and the thrusting of one layer over another, plus the eroding effects of glaciers, water, and wind. The Appalachians in eastern North America are also the result of a plate collision, but a much more ancient one (starting around 300 mya).

The second major mountain forming process is volcanism. Volcanic mountains can be isolated, such as Mount Kilimanjaro in Africa, or can be found in ranges, such as the Andes on the Pacific rim of South America (which is also a site of relative tectonic plate movement). Due to their formation process—molten lava being brought to the Earth's surface—volcanic mountains tend to have a form different

[1] The classical source for the SE account of functions is Wright, L. (1976). However, Wright's account is independent of any particular biological theory of etiology. Brandon (1981) offers the first SE account of biological function explicitly tied to modern evolutionary theory.

[2] The classical source for the CR account is Cummins (1975). But Amundson and Lauder (1994) is the best source for applying this account in biology.

from those produced by the folding and thrusting of Earth crust. Volcanic mountains tend to be conical and tend to be more isolated from each other in contrast to the long ridges with connected peaks often formed by continental collisions.

The term "mountain" has some vagueness attached to it. How tall does a hill have to be to be a mountain? How do we distinguish between multiple peaks of a single mountain and multiple mountains? Geologists do make these distinctions operational, but in a somewhat arbitrary way. This is not our concern. Rather our concern is with the difference between historical/etiological concepts vs. ahistorical ones. The concept of mountain is ahistorical. So is that of a mountain ridge. Likewise for that of a conical mountain. The formation history of these geological entities (their *orogeny*) is not a part of these concepts. An easy way to see this is to ask what would happen to them were our theories of orogenesis to be radically overthrown. Surely, these ahistorical concepts would survive.

In contrast, the concept of a volcanic mountain is a historical one. Something counts as a volcanic mountain if and only if it has a certain causal history. Similarly we could use the rather inelegant term "continental-collisional" to describe mountains that have an orogenesis in the collision of continental plates. One need not be an expert geologist to apply these causal-historical concepts to certain mountains. Mt. Kilimanjaro is clearly volcanic. Mt. Everest is clearly continental-collisional. But for more ancient mountains, we nonexperts might well go wrong in trying to apply these terms. For instance, in the Massif Central in south central France, there are both volcanic and nonvolcanic peaks. Due to weathering, it is not always obvious which is which. (Is Mount Aigoual volcanic? No, its granitic underpinnings are inconsistent with a volcanic origin.)

This raises an important point about causal-historical concepts. They are epistemologically riskier than their ahistorical counterparts. One might use this fact to argue for a sort of conceptual monism—a banning of historical concepts. We will discuss some arguments for conceptual monism in the next section. For now we need only note that no geologist would take them seriously. Here is why.

Consider the following generalization: Most volcanic mountains are cone-shaped, and most cone-shaped mountains are volcanic. Although this is not the most scintillating of geological generalizations, it is true. And to state it, we need both ahistorical concepts (cone-shaped) and historical ones (volcanic). The knowledge stated by this generalization cuts across both categories (historical and ahistorical). Why would we adopt a conceptual scheme that makes impossible the statement of such generalizations?

Not all such generalizations are geologically trivial. Consider this: There are over 100 mountains on Earth that are over 7,000 m in elevation above sea level. All of these mountains have plate tectonics as their orogeny. Put another way, none of these mountains are volcanic. The highest volcano on this planet is Ojos del Salado on the Chile-Argentina border. Its elevation is 6,893 m. This fact—put briefly, that all of the highest mountains on this planet are continental-collisional—might strike one as a paradigm of what philosophers of science call an accidental generalization, that is, one that happens to be true, but that has no law-like force behind it. However, that is probably mistaken.

Two main factors limit the maximum height of a mountain on a rocky planet: (1) the mass of the planet and hence its gravity and (2) the strength/stability of its crust. The higher a mountain peak, the more massive the mountain—indeed because of scaling effects, the mass will rise exponentially with height. Mt. Everest is probably pushing the maximal height a mountain can achieve on Earth due to the principle of isostasy, which describes the balance of the buoyancy of the uplifted continental crust over the denser mantle. Physical force is required to counter the gravitational pull of the planet. Volcanic activity probably cannot match the energy of plate tectonic activity. If all this is true, then it is no accident that all of the highest mountains on Earth are nonvolcanic in origins. We can use Steve Gould's (1989) analogy of the tape of life here: Were we to rerun the physical evolution of this planet over and over, most of the runs would be one where our generalization about tallest mountains comes out true.

(In support of the above, it should be noted that the highest known mountain in our solar system is Olympus Mons on Mars. It has an elevation of 27 km above the mean surface level of Mars, which is over three times the elevation of Mt. Everest. It is volcanic. But Mars is much less massive than Earth, and its crust is more stable—there is no known plate tectonic activity on Mars. Thus that a Mars-like planet would have a higher volcano than an Earth-like planet is predictable.)

What we have seen in this brief excursion into the geology of mountains is that basic generalizations and interesting hypotheses require a conceptual scheme rich in both historical and ahistorical concepts. There is no good reason to impoverish such a scheme in geology. As we will see in the next section, the same is true in biology.

2 The Analogous Situation in Biology

Biology, like geology, is a historical science. But the situation in biology is more complicated for two reasons: the first epistemological and the second conceptual. As we will discuss shortly, historical concepts in biology are epistemically riskier than similar concepts in geology. On the conceptual side, there has long been conceptual confusion, and sometimes outright conflation, of concepts going by the name "adaptation." So before turning directly to SE and CR analyses of function in biology, let us consider the closely related concepts of adaptive traits and traits that are adaptations.

The mainstream view in evolutionary biology is that to say of some trait that it is an adaptation is to attribute to it a particular causal history (see, e.g., Brandon 1990). Briefly, a trait is an adaptation if and only if it owes its population frequency and distribution (mainly) to the process of evolution by natural selection. But not all biologists have agreed on this, and the primary reason for dissent has been the epistemological riskiness of the historical concept of adaptation. Unlike the concept of a volcanic mountain, which has some epistemic risks, but none that would trouble an even mildly competent geologist, this historical concept of adaptation is quite difficult to apply (see Brandon 1990, chap. 5 for a characterization of what is required of an ideal adaptation explanation). Thus, Bock and von Walhert (1965) in an influential paper argued for a concept of adaptation that was based purely on current effects, not on history. More recently, Reeve and Sherman (1993) argued for an exclusively ahistorical definition of adaptation. They characterize an adaptation as a trait that has the highest fitness among currently available alternative traits in the current environment. There is no troubling history here.

But the costs of such a move are excessive. First, nothing is added to our conceptual repertoire by this move. Adaptation, in the Bock and von Wahlert and Reeve and Sherman sense, is equivalent to calling a trait adaptive in its current environment. That notion, which relies on engineering/ecological analysis (CR analysis), is certainly important in evolutionary biology. But we already have that notion. Their proposal is one of conceptual contraction. It is one thing to recommend caution in applying epistemically risky concepts, but quite another to ban them altogether. Only excessive epistemic timidity would call for this. But if that is our motivation, why not become a Berkeleyian idealist and risk nothing?

Restricting one's conceptual repertoire to only one side of the historical/ahistorical divide would, as in geology, make impossible certain generalizations, hypotheses, and questions. For instance, are most adaptive features adaptations? Are most adaptations still currently adaptive? Are functional shifts common, that is, do most adaptations still selected for because of their initial function, or has there been a functional shift so that what once was an adaptation for *X* is now an adaptation for *Y?* Each of these questions is important and the subject of current controversy. But if the questions are important, they need to be meaningful, and they are not meaningful if we adopt the impoverished conceptual scheme recommended by Reeve and Sherman. (Of course, these questions would be meaningless in a conceptual scheme populated only by historical concepts, but no one that I know of has recommended that.)

Are most adaptive features adaptations? It was precisely this question that led Gould and Vrba (1982) to campaign for the addition of the term "exaptation" to our evolutionary vocabulary. Although that term has not caught on, and I think it is unnecessary, their conceptual point is certainly correct. And that point is that we cannot rule out *a priori* the existence of traits serving some current function that arose purely by chance, nor should we rule out traits that did evolve by natural selection that have been co-opted for a new use. Consider the latter first. Gould and Vrba (1982) suggest feathers as a plausible case of this. It looks like feathers originally were selected in nonflying dinosaurs for their thermoregulatory effect. Feathers are good insulators. But later, in early flying dinosaurs (especially those that became modern birds), feathers were presumably selected for their use in flying. If this is correct, then at least during the early evolution of bird flight, feathers were adaptive in their aerodynamic effects, but were not adaptations for flight. (Another good example of this is the case of insect wings, discussed below.)

Gould and Vrba's (1982) first case of possible adaptive non-adaptations is even more interesting. We now have a very good, and plausibly very general, example that illustrates this point. Consider so-called pseudo-genes. Pseudo-genes result from a duplication event in the genome resulting in two or more copies of a gene where once there was only one. These new copies typically serve no function (i.e., they literally do nothing but take up space). They, the new copies, are not (organismic) adaptations. They may, and probably should, be thought of as gene level adaptations in that this process of duplication and reinserting elsewhere in the genome is a form of gene level selection. But in this case, what is adaptive at the genic level is at best neutral at the organismic level. These bits of DNA are now free to change by mutation, that is, selection no longer constrains, by eliminating, mutant version of the gene. Most such mutants will cease to be transcribed and will be functionless. But by chance, one of these mutant genes may take on a new function. This has been termed "neofunctionalization" in the molecular evolution literature and is thought to be an important source of evolutionary novelty (Lynch 2007a, b). Whether or not that is correct is not the issue here. Here, the point is rather simple: Do not adopt a conceptual scheme that precludes the investigation of this sort of evolutionary phenomena.

Are most adaptations still adaptive? Perhaps the answer to this is yes, but there are certainly exceptions, for example, the wings in Emus. Although Emus do not fly, their wings may well have been co-opted for some other use and subsequently molded by selection for that new use. That then would be a case of functional shift, discussed next. But again, we should not rule out *a priori* the existence of purely function-free vestigial traits. (Does the human appendix really serve an evolutionary function?)

Are most adaptations still adaptive for the same function that drove their initial evolution? Here, we are asking how common are functional shifts in evolution. They certainly occur. A good candidate for such a shift is the evolution of insect wings, which was initially driven by the adaptive advantages of short proto-wings for thermoregulation and only later driven by the adaptive advantages of flight (Kingsolver and Koehl 1985). To study this interesting phenomenon, we need a conceptual repertoire rich enough to describe it. Restricting our repertoire here would be as foolish as it would be in geology.

If the reader agrees with all of this, then the case has been made for pluralism with respect to functional analyses in biology. On the ahistorical side, we have the concept of adaptiveness. To say that a trait is adaptive is to say something about its causal role in the here and now. Briefly, it is to say that the trait has a CR function (connected with the fitness of the organism(s) possessing the trait, in the current environment). Thus, unless we want to give up this concept, we are committed to CR functional analyses.

On the historical side, we have the concept of adaptation. Adaptation is a concept relating to causal history. A trait is an adaptation only if it has an SE function.

Furthermore, neither type of functional analysis is reducible to the other. SE functional analysis is dependent on CR function in the following way: to say that a trait has an SE function is to say that some past CR function led to the origins/maintenance of the current character state. In this way, CR functional analyses are more fundamental than SE. But that does not mean that the SE account reduces to the CR and so can be eliminated. The SE account implicates a particular sort of causal history, one that involves a CR function, but much else as well.[3]

---

[3] See Brandon (1990), chap. 5 for a detailed account.

3 Form, History, and Function

The main argument of the last section is that biology needs the causal-historical concept of adaptation as well as the ahistorical causal concept of current adaptiveness in the current environment. That is, we need to talk about adaptations in the strict historical sense, as well as adaptive traits, to even frame the questions briefly explored above. How does this bear on the analysis of function? The answer to this has already been hinted at in Sect. 2, but I will now make that explicit. To talk of adaptive traits is to engage in CR functional analysis. The wings of most birds function as flight mechanisms. Not so for Emus' wings. But if they are adaptive in some other way, say they are used in mating, then that could only be confirmed by an ecological/engineering analysis. This just is a CR function.

Adaptations, in the strict historical sense, are things with SE functions. This is not controversial in the least. We have seen that some would try to eliminate this concept entirely, but we have also seen that it would be foolish to do so. Thus, SE functions are firmly embedded in our evolutionary conceptual repertoire.

4 Conclusion

Pluralism with respect to functional analyses in biology is a special case of a more general conceptual pluralism of historical and ahistorical concepts. In geology, we saw that important questions, hypotheses, and generalizations can only be formulated within a conceptual repertoire that contains both sorts of concepts. The situation is exactly analogous in biology. We need to talk both about adaptive traits and adaptations. We need to talk about CR functions and SE functions. Thus, functional monism in biology is unsupportable.

References

Amundson, R., and G.V. Lauder. 1994. Function without purpose: The uses of casual rose function in evolutionary biology. *Biology and Philosophy* 9: 443–469.

Bock, W., and G. von Walhert. 1965. Adaptation and the form-function complex. *Evolution* 10: 269–299.

Brandon, R.N. 1981. Biological teleology: Questions and explanations. *Studies in History and Philosophy of Science* 12: 91–105.

Brandon, R.N. 1990. *Adaptation and environment*. Princeton: Princeton University Press.

Cummins, R. 1975. Functional analysis. *Journal of Philosophy* 72: 741–765.

Gould, S.J. 1989. *Wonderful life: The Burgess Shale and the nature of history*. New York: Norton.

Gould, S.J., and E.S. Vrba. 1982. Exaptation–A missing term in the science of form. *Paleobiology* 8(1): 4–15.

Kingsolver, J.G., and M.A.R. Koehl. 1985. Aerodynamics, thermoregulation, and the evolution of insect wings: Differential scaling and evolutionary change. *Evolution* 39: 488–504.

Lynch, M. 2007a. The frailty of adaptive hypotheses for the origins of organismal complexity. *Proceedings of the National Academy of Sciences USA* 104: 8597–8604.
Lynch, M. 2007b. *The origins of genome architecture*. Sunderland: Sinauer Associates.
Reeve, H.K., and P.W. Sherman. 1993. Adaptation and the goals of evolutionary research. *The Quarterly Review of Biology* 68(1): 1–32.
Wright, L. 1976. *Teleological explanations*. Berkeley: University of California Press.

Weak Realism in the Etiological Theory of Functions

Philippe Huneman

Abstract The etiological theory of functions advocates a realist view of functions, through a construal of functional ascriptions as statements about evolutionary history. Basing functions on fitness and natural selection, it faces difficulties when it comes to discriminating between distinct Equal-fitness properties of one trait. I argue that evolutionary theory alone cannot justify a fine-grained determination of what is the function of the trait in those cases. Biologists have then to choose a specific method to establish the nature of the function; three methods (a counterfactual one, a comparative one and one oriented towards organismal organisation), each committed to specific explananda, are here studied, with examples. They may yield distinct functional ascriptions for the same trait, which introduces an element of explanatory dependence within the etiological account of functions.

Since Millikan (1984) and Neander (1991a, b), a growing consensus on a theory of function has arisen that would account at least for the uses of functional ascriptions and explanations in all sectors of what Mayr called "evolutionary biology" or biology of the ultimate causes (Mayr 1961).[1] In all its versions, the etiological theory assumes a realist stance towards functions: those are not only elements of our descriptions of nature; they exist in the biological and ecological domains. Functional properties, such as the predator-repelling effect of the eyelike spots on some insects' wings, are as objective as other properties like mass. In this respect, etiological theories differed significantly from the causal role theories of function formulated first by Cummins (1975), which conceived of functions as relative to systems that are chosen and delimited according to the explanatory interests of the scientists.

---

[1] See comments in Beatty (1994) and Ariew (2003).

P. Huneman (✉)
IHPST (CNRS/Université Paris I Sorbonne), Paris, France
e-mail: philippe.huneman@gmail.com

P. Huneman (ed.), *Functions: selection and mechanisms*, Synthese Library, Studies in Epistemology, Logic, Methodology, and Philosophy of Science 363,
DOI 10.1007/978-94-007-5304-4_7,

In this chapter, I claim that if one wants to account for the genuine biological notion of function, one cannot entirely endorse this realist stance, since functional ascriptions by biologists display, when they are for fine-grained functions, a dimension of explanatory dependence. This implies that if one wants a general theory of function that accounts for every functional attribution occurring in one evolutionary domain or other, one must be committed to a weak realism, meaning that while the coarse-grained determinations of functions are objective, the peculiarities of those functions are somehow explanatorily dependent.

The first section presents the requirements for an etiological theory of functions, that is, requirements of being explanatory, normative, realist and discriminative; the second section presents the discrimination problem that those theories face, a problem raised by the multiplicity of properties necessarily related to a property that is shown to have been selected and is then a candidate for being the function of a trait. The last section argues that in order to face those issues, one must choose some methods to discriminate between candidate functions and that those methods involve specific explananda for functional explanations based on functional ascription, so that they ultimately yield divergent fine-grained functional ascriptions.

1 The Etiological Theory as a Realist Theory of Functions and Its Requisites

Roughly put, the etiological thesis says:

> Function of trait X is Z if X has been selected for doing Z.

In Neander's language, F is then a "selected-effect" (SE) function. This is a claim about the *meaning* of functional ascriptions. Before entering further analysis, I have to make two points clear, one about the scope and one about the intended consequences of this thesis.

The scope: it is not obvious whether this analysis holds for all the functional discourses, thereby for functions of artefacts or only for biology. Some authors like Wright (1973)—even if his theory was not exactly an SE theory—or Millikan (1984) have devised a general theory, whereas other authors such as Neander (1991a, b), Godfrey-Smith (1994) or Vermaas and Houkes (2003) restrict the scope of the analysis to the biological domain. One difficulty, if the theory is intended to widen the scope up to the whole functional discourse, is obviously that "selection" is well defined only in biology as *natural selection*. Its equivalent concerning non-biological objects is difficult to define, first of all because the relationship of inheritance between entities is not understood concerning artefacts or institutions as well as it is in the case of organisms.[2] Millikan (1989) conceptually defined a relation of "copying"

[2] See Lewens (2004) on the shortcomings of an etiological theory for culture or artefacts.

supposed to obtain in all relevant cases and of which genes would be a particular case, but this requires specific theoretical work that is unlikely to be unanimously accepted. So here I will take the etiological theory as an SE theory valid in the case of biological discourse and leave the case of its extension undecided.

Now, what is the etiological theorist aiming at? This sort of query is of general concern in philosophy of science, but here it is particularly important because it touches on the content of the analysis. Grossly said, a claim in the philosophy of science can be either descriptive or normative. When it's descriptive, it provides an analysis of the concepts and their uses by scientists. When it's normative, it provides an analysis of the concept that is meant to be a criterion according to which one can distinguish between the right and wrong uses of the concept by scientists. Of course, the boundary is not so clear, since sometimes the result of a conceptual analysis means that some of the uses that are at odds with it too much turn out to be misguided, and, reciprocally, a normative analysis providing a concept that fits none of the actual uses by the scientists would not belong to philosophy of science.

However, in the case of the etiological theory, some authors like Millikan intended to forge a "theoretical definition" of function; this definition is fulfilled by a lot of the current uses familiar to evolutionary biologists, but it's not her primary aim to capture biological usage. Her theory is actually a general naturalistic theory of intentionality and language. Such a theoretical definition does not have to correspond adequately to all the biologist's uses. This also implies that the scope of Millikan's analysis will be wider than biology. On the other hand, Wright's analysis, while also wide, is a conceptual one, in the manner of Strawson's descriptive metaphysics, trying to unpack what we say in general when we say "the function of Y is…". This conceptual analysis relying on the usual utterances of "function" contrasts with Neander's (1991a, b), Godfrey-Smith's (1994) or Buller's (1999) analyses, which mostly focus on scientific uses of the concept of function. Those authors restrict their scope to biology. The conceptual analysis must not be at odds with actual practices of biologists, and too much of a gap between its results and some of those practices would be a decisive objection. All the uses of the functional concepts and the functional explanations should not be assumed to be maximally coherent and consistent; therefore, it is to be expected that the philosophical analysis will prove some of them to be mistaken. However, at least some of the general features of functional discourse in biology should be captured by those analyses, and more generally, those features should stand as evidence in the debates about conceptual analyses of functional discourse.

So here I take the etiological theory as a *conceptual analysis of scientific discourse in biology*, assuming that no concept of function can make sense if it is not firmly connected to the scientist's practice of functional explanation and functional attribution. Yet, there is another ambiguity, hence a second distinction to be emphasised: conceptual analysis addresses functional *ascription*—as Wright says, it is intended to unpack what "X is the function of Y" means. If Neander (1991a, b) is right, X is a selected effect that explains why Y is here. But what about the way this function of Y was established? Conceptual analysis here shares the fate of Frege's analysis of arithmetic: while it unveils what it is to be a number, it says nothing about how we can have access to numbers and what we do with them, that is, counting. But science cares about accessing such objects, so if the conceptual analysis is relevant to philosophy of science, it should say something about how this function of Y can be known.

Most of the time, there are several means through which such function of Y can be known, and they are embedded in distinct *functional explanations*. The rest of this chapter will develop this claim, but for the moment, let's expose the distinction between the functional *ascriptions*, captured by the etiological theory, and the modes of functional *explanation*. Saying that the function of morning sickness is to filter the toxins likely to harm the fetus, since its defences against toxins are far less developed than the mother's defences, somehow explains *morning sickness* (Nesse and Williams 1995; Profet 1992). However, by "functional explanation", we could also think of the explanation *of the defence system of the fetus* by the mother's nausea. If we ask "how does the fetus protect her/himself from toxins during the period in which it is the most vulnerable?" we may cite, among the defences, the mother's morning sickness, the intensity of which decreases after the first month in exactly the same way as the fetus's vulnerability decreases. So while (according to the etiological theory) functional *ascriptions* explain the presence of the functional item, some functional *explanations* explain the functioning of a general system, of which the functional item is a part, through precisely the function identified in the ascription. The standard view would say that those second explanations pertain to another concept, a causal role concept of function (Cummins 1975) according to which the function of X in the system S is which contribution X makes to the functioning of S (Godfrey-Smith 1994; Millikan 2002). However, my point here is that the functioning of S in general is *identified* through the SE functional ascription; for example, without the description of the morning sickness as a filter system (which is an evolutionary answer to a question about the presence of an apparently maladaptive trait and therefore a clear case of SE function), the functional explanation of the fetus's defence system would be incomplete, first of all because delimiting the fetus's system in general (i.e. with the inclusion of the mother) would not be possible. In other words, without the ascription of the SE function, the system of which the morning sickness is a part wouldn't be identified; thus, the explanation in terms of Cummins functions could not be stated. (Notice that this example is not the only mode of functional *explanation*.) Therefore, the etiological theory also has to account for the *explanatory uses of SE functions*, which is something more than the general analysis of functional ascriptions. Those uses, as we will see, will be my reason for casting a doubt on the realism of this theory.

But first let's remind ourselves of the general features of such theory—or the requisites for any theory that would pretend to be an etiological theory of functions:

1. *Explanatory*. It accounts for the *explanatory force* of functional ascriptions: those explain the presence of the item. This is the main difference from systemic accounts of functions *à la* Cummins: in the latter, functional ascriptions just contribute to the explanation of a general capacity of a system, while in the former, functional ascriptions *are* explanations of the presence (maintenance, origin or special location) of an item. The ascription is thus explanatory *by itself*.

2. *Normative*. It accounts for the *normative character* of functions. In effect, we can make sense of cases where we say that the function of X is Y but X cannot do Y. This is implicitly a normative claim: it means that X is abnormal. In the light of SE theory, normative claims make sense of a statement of this sort: "abnormal instances of function Y are tokens of the type X (which is said to have the function Y because its prior tokens have been selected in the past for doing Y) that are themselves unable to do Y". This aspect of selected-effect theory is irrelevant to us for the moment.
3. *Realist*. It is a *realist* theory. Yet, what is this realism about? It means that those theorists hold that "function" is a legitimate and not a replaceable concept of scientific explanation, not a shorthand for something else, and that all attempts to reduce functions to some kind of nomothetic, effect-cause relationships, as initiated by Nagel (1961), are misguided. To be a function is a natural property of some items, not an epistemological characterisation of them. So, first, we cannot reduce functional terms and functional discourse to another kind of discourse, in terms of non-functional categories (such as causal relationships between categorical properties). And, second, functions are independent from our chosen explanatory interests and strategies: they are something real in nature (a new kind of property, not something to be reduced to covering laws and categorical properties, etc.).
4. The third aspect entails a major consequence: SE theory is *discriminative*, in the sense that it accounts for the difference between the accidental effects of a trait and effects that are "the functions of the trait". Think of the most famous example of the heart: its making of noise is an accident, but circulating the blood is its function. This thus makes a *real and genuine difference* between two classes of effects. Here, SE theory contrasts with systemically inclined theories of function: if the function of X is its proper contribution (or capacity to contribute) to the functioning of a system defined by us, change the system and you will change the function. For instance, while circulating blood is the function of the heart in some systems (more or less akin to "organism"), making noise is a function in another system (the patient-physician dyad). So the *discriminative* feature is proper to SE theory.

Now, *why* is SE theory realist? The main reason is that functions are traced back to the *causal history* of the item; ascribing functions makes a statement about this causal history (Millikan 1989); hence, it is not dependent on our interests because the causal history is mind independent. (This is a very general claim, and the meaning of "causal history" differs according to different authors; we leave aside the debates on the determination of this history: Does function account for the maintenance of traits, i.e. history is only recent history (Godfrey-Smith 1994)? Is this causal story mainly a "reproductive history", where cause is loosely defined and mixes a production and a copying relation, like in Millikan (1984)? In any case, interpreting functional ascriptions in terms of causal history is a reason for realism about functions.) On this basis, what does justify the *discriminative* claim? Why are functions *really* different from accidents (side effects)? Because they are the only properties for which the trait has been selected. There has been no selection for side effects or accidents, even if those effects have been selected in the same time with the traits. Side effects are not the reason of the selection: we can think of a possible world where X, which in the actual world has function Y and side effect Y′, is split into two entities, X and X′ with respective effects Y and Y′; in such a world, X′, unlike X, would not be selected. Hence, Y′ is not the function of X. It appears here, as noticed by Millikan (1989) and Agar (1993), that counterfactual statements lie at the basis of the *discriminative* thesis. So, briefly said, why does SE theory have the characteristics of being discriminative? Here, *fitness* does the job: it accounts for a real difference between several effects and properties of a trait (Walsh 1996). The function of the heart is to circulate the blood because this property *is fitness enhancing*. Among several effects of trait T, some never enhanced the fitness of the organisms in its context, so they are not the function of T, but mere by-products, accidental side effects, and in the counterfactual world described above, they would not enhance fitness and would not have been selected.

2 The Weaknesses of SE

Taken together, the above requisites face several problems. The literature on it is vast and I won't be exhaustive of course. I will just focus on some problems mostly raised by requisites 3 and 4.

2.1 Logical-Type Problem

SE theories ground the discriminative power of functional ascriptions on fitness (Walsh 1996). But fitness of trait T is a relational, *comparative*, *quantitative* concept: (according to the usual notations) w(T) changes according to the environments, and biologists use generally relative finesses—the availability of absolute finesses is not necessary for biology; on the other hand, function is a *categorical*, *qualitative* concept (T has, or does not have, the function Z, and this does not seem relative to other organisms having traits with this function (or not)…).[3] Fitness and function prima facie don't seem to belong to the same logical kind.

[3] For a suggestion about items being more or less functional, this being grounded on a theory of functions, see Wimsatt (2002) and this volume. One could say here that fitness does not change according to the environments because it can be seen as a dispositional property that correlates to each environment a proper number of offspring or probability distribution of offspring. But this does not solve the main problem, which is the quantitative nature of fitness compared to the qualitative nature of function.

This is not a major problem since there is a way of overcoming the difficulty by rephrasing the dependence of functional ascriptions upon the concept of fitness. What yields the ascription is not the value of the fitness of the trait (which is relative to the fitness of other traits or other organisms bearing variant traits), but a fact: "the fact that trait T in context C had fitness "n" because it did Z". This fact, as such, is true or false; it's not quantitative, and it is not a measure. This is what is involved in causal history to which the functional ascription refers. To this extent, such a fact is homogeneous to a functional ascription—it is nonrelative and nonquantitative—hence, there is no logical discrepancy between fitness as relative and quantitative and function as a categorical property. A corollary of this solution is that when it comes to a reference to "causal history", the etiological theory must rely on a conception of causation as linking facts rather than events. This immediately leads to our second problem.

2.2 Problem of the Bundle of Effects

This issue is akin to a problem sometimes called the determination of content (Dretske 1986; Neander 1990, 1995; Agar 1993; Millikan 1989; Enç 2002; Price 1998), mostly in relation to the naturalisation of semantics. In the present context, we can formulate it in the following way: there is not *one* series of facts to be related in a causal history. "Running faster than 25 mph" does not mean "running faster than a tiger", so those statements refer to *two* different facts—even if they are about the same event.[4] More generally, there are several facts leading to different but equally legitimate functional ascriptions because they all concern properties of organisms that have the same fitness. Therefore, contrary to the above claim intrinsic to SE theory, selection and fitness *alone* cannot discriminate enough between properties of a trait T. A bundle of effects can compete to be the true functional ascription.

How can this bundle of effects be manifested?

On the one hand, the properties may be included in each other: to run, to run at 60 mph, to run to escape a lion, etc. are nested facts. On the other hand, they may be independent: to retain heat and to hide from prey are both properties of the fur of the polar bear and reasons for its selection, but they can occur separately (the first is about the texture, the second about the colour of the fur).[5] Maybe this second case is not so difficult, as we can admit two functions here because we can discriminate two selective pressures or environmental demands bearing on the same trait; eventually

---

[4] Suppose you have the event of a white tiger running at 30 mph in the tundra. This is ontologically a single event; however, it can be picked up by many propositions ("white tiger running", "white tiger running at 30 mph in Siberia", etc.), each of those having a proper meaning. All those meanings correspond to "facts".

[5] This raises the huge issue of combining selective pressures, whereas this combination is not additive. Such question involves issues about the causal nature of natural selection (see Lewens 2009) which are not directly under focus here.

the fitness value of the trait as such is a trade-off between both. But concerning the first case, the only one of importance here, let's notice a crucial distinction: some inclusions are *entailments*, necessary relationships, like "to run at 60 mph" entails "to run"; other properties are *contingently related* in a context-dependent manner: it is context dependent that moving dots are flies or that running faster than 60 mph means faster *than lions*. The bundle of effects, here, occurs in *our actual* world (as compared to other possible worlds).

Now, one could object: where is the problem? Why not allow multiple attributions (i.e. all of those properties, being equal in fitness, are the functions of trait Z)? I can conceive of two equally important reasons to reject this option:

- Firstly, to run faster than lions implies to escape predators, which implies to enhance survival; in the end, all functions of all traits of all organisms would be "to enhance survival", which is absurd. (This is perfectly developed in Enç (2002) under the name of "*landslide* argument").
- Secondly, even if we do not go entirely down this landslide route (e.g. by conventionally deciding that "yes, all functional traits have one function in common, but let's keep this universal function aside"), several different items would still have one property in common and then might have the same function, which goes against the idea that an item X must have its proper function: ears and eyes have the property of helping flight from predators—hence, they have the same function—but biology does not use such an indiscriminate concept of function.

Thus, fitness discriminates several effects of a trait *as a whole* which are either logically entailed by one another or contingently connected in a context-dependent manner. So we need a criterion to establish *the function* of trait T because selection and evolutionary history cannot provide it by themselves. Therefore, we should consider the *methods of establishing* the function of T; this will lead us to consider the kinds of functional *explanations* that are using the functional *ascriptions*, in which SE theory is exclusively interested. And nothing ensures us a priori that those methods yield the same result.

3 Establish and Explain Functions

I distinguish three methods here, for our purposes, and will define them in the course of the analysis: (a) functional organisation schema, (b) counterfactual design analysis and (c) comparative analysis. Those three methods are able to deal with the "bundle of effects" problem in such a way that, functional ascriptions being made discriminatory, they can enter into genuine specific explanations. Their common feature, which solves this problem, is that *explanation at a level other than the trait* allows ascriptions to be discriminatory at a finer grain than mere fitness does, as we will see below. There are three ways of establishing functions by undertaking an explanation of some specific explanandum.

3.1 *Functional Organisation Schema*

Here, the functional ascription enters into a reconstruction of a general system of nested functions (which more or less maps onto an organism). The trait may be included in an organised system, such as that which would be a common function of two traits is ascribed to a more general system. In the end, all traits have one function because a function shared by several traits is then ascribed to an upper-level trait. For instance, "hear noises" is the function of the ears, "see moving shapes" is the function of the eyes, "detecting predators" is the function of a more general sensory device and so on. The main assumption is (D) if two traits have one function in common, this function should belong to a more general system of which they are subsystems. Here, the function of trait X is specified, beyond the fitness-grounded SE characterisation, by nesting X into a general system in order to explain the functioning of the extant organism. The schema of a functional organisation in Wimsatt (2002) gives an idea of what it's all about. Functional organisation looks like a descending tree, and two traits having a candidate SE function X in common are related to a higher-level trait to which will be ascribed this X as an SE function, according to (D). Nodes are traits that have a function which is common to the traits related to them. Wimsatt claims that functional organisations are rarely exact trees, because of the frequent functional loops (2002, 187), and also that they are made by activities rather than entities in order to avoid multiplication of paths between nodes (2002, 183): those considerations, however, are external to the argument developed here, which would still subsist with some modifications in Wimsatt's framework.

So concerning the "bundle of effects" problem, the discrimination is in fact done when one tries to explain how all those traits are put together to work and bring about a functioning, reproducing and surviving organism, because (D) has to be assumed to be a principle of this investigation. In fact, when the ascription of a function to trait T on the basis of SE theory is ambiguous, if we compare T to other traits in a same organism and then draw a functional schema with nodes, according to (D), the problem disappears because the candidate functions of T will be distributed along the nodes on one or several paths. In this respect, SE functions are in fact complemented by causal role functions: it is only by considering the organism as a system, thus distinguishing its contributing capacities, that one can consider trait T in its difference with trait T′ when both have been selected for some common reasons, and then specify what is proper to T in its contribution to the organism's functioning.[6]

As a real life example, I will use here the evolutionary studies on symmorphosis, namely, the hypothesis that the organs are optimised together regarding all of the environmental demands bearing on an organism. In a study on the design of nerve fibres, Keynes (1998) writes: "the first and overriding respect in which the structure

[6] On this hierarchy and the relation between functions of parts and function of their whole, see Huneman (2007).

of peripheral nerves has been optimized lies in digitalization, ensuring that the information conveyed depends on the pattern and number of impulses transmitted by each fibre, and it is not at the mercy of conduction time that might vary from time to time with local conditions. The second respect is that the size of myelinisation of fibres is closely adapted to their specific function so that the largest ones are preserved for pathways where high speed of conduction is essential, and the smallest and slowest ones are used for sensory pathways where rapidity is not a primary requirement, or for control of the autonomous nervous system" (276). This, in turn, is easily interpreted in the schema of functional organisation. "Myelinisation of fibres", in some pathways like motor pathways, has the function of *quickly* conducting signals; in the sensory pathways, it has the function of *slowly* carrying it. The more general trait "structure of peripheral nerves", encompassing those two items, has the general function of digitalising information in order to make it robustly deliverable. So with some adjustments, we see that the two connected putative functions, *digitalising information* and *conveying it* at some relative speed, can be disambiguated as candidate functions of the structure of peripheral nerves according to the SE theory.

3.2 Design Counterfactual Analysis

3.2.1 The Simple Case

Another question to be addressed on the basis of a coarse-grained ambiguous SE functional ascription is the following: "we have the trait within the system; given the bundle of effects here considered, what is the problem solved by this trait *that is better solved in this way* than in another way ?" The idea of "reverse engineering"[7] is similar to this method. So we have to define in which regard (i.e. according to which of the effects included in the bundle of effects) the trait is optimal, with respect to other possible traits. This is a counterfactual analysis, since the variants considered here don't need to exist.[8] A mathematical optimality model can do the job perfectly: variants are the values of some variables, and even if some values, sets of values or combinations of sets of values of some variables never existed, we can still plot the fitness values of the traits so defined in the model. Then we will find out the relevant variables with respect to which trait X is optimal.

Let's first consider a famous toy case. We want to know the function of the dots-fly-catching devices in a frog—is it to catch moving dots or to catch flies? Suppose we have animals X that are tracking dots but not flies and animals Y that are tracking flies but not dots. There are two variables, fly-sensitivity F and dot-sensitivity D,

[7] See Lewens (2004) for an analysis.

[8] Wouters (2003) also considered those kinds of counterfactual analysis and connected them to the idea of design. However, we are addressing here a quite different question.

with binary values. X is such that D = 1, F = 0, and Y is such that D = 0, F = 1. In a given environment E like the frog's real one, moving dots are flies. On the basis of mere fitness, we can't say whether the function of the device in the frogs in E is to detect dots or flies because in E, the fitness of X equals the fitness of Y. But in another environment E′ where flies are not dots, all things being equal, Y will have a higher fitness than X since it will get preys. The only relevant variable is then F, and the function of the device is to catch flies.

From this perspective, we always ascribe "ultimate" against "proximate" functional content (in the terms of Horan 1989) or "benefit" function instead of "stimulus" function (in the terms of Neander 1990). This is due to the fact that if the stimulus and the benefit are contingently connected in the actual environment so that they define traits with the same fitness, in other possible environments, they are not connected, and hence by definition, the trait determined by the benefit is the only one that is selected.

The counterfactual method here is absolutely natural since we have to disambiguate effects of a trait that are *contingently* related, which means specifically that in some other possible worlds (more precisely here, in some other environments), they are not. Of course, there would be no point using this method to discriminate putative functional ascriptions when the candidates are *necessarily* related; in such a case, one should resort to the functional organisation schema.

Notice that in this method, there are two levels of counterfactuals: the first level bears on the trait (considering devices where the values of the two variables describing the traits, that are equal in this world, are different), and the second bears on the environment (considering environments where both properties are indeed different). Millikan (1989) objected to a counterfactual approach of functional ascriptions because she said that counterfactuals were indeterminate if it is only a case of having versus not having trait T ("not having" is indeterminate). However, in fact those counterfactuals defined variables by traits, and environment variables are perfectly determined since we can define them in terms of world specified only according to the values of the variables.

Now, what here is explanatory? Saying that the relevant variable is "fly tracking" means that the effect of the trait "detector device" that *explains its potential maintenance* in a population, no matter what the actual history of this population has been, is its fly-tracking ability, rather than all others, because only this one accounts for the fact that the trait is likely to be selected. The method therefore aims at explaining what the environmental problem that the trait has been selected for solving is, hence the problem regarding which extant organisms are optimised. So the functional ascription explains part of the general design of the organism, provided that to be designed means to be in some respects optimal, or more exactly, resulting from trade-offs between divergent optimality requisites (Stearns 1992). To this extent, the method sketched here—namely, looking for the problems actually solved by the traits in the environment of the organisms—therefore aims at uncovering the general design of the organism as a set of interrelated optimised devices or problem solvers. In other words, the functional ascription here enters into a general explanation of the way the organism is dealing with its environment. Undertaking such an explanation compels one to adopt the method sketched here in order to attribute functions to traits when they raise a "bundle of effects" problem. For this reason, I called the method "counterfactual design analysis" because, while it resorts to counterfactuals, it uses them with the background of a general assumption about design (meaning that organisms can be conceived of as integrated problem solvers[9]).

3.2.2 More Complicated Cases

Of course, the case sketched here has been simplified significantly. Most of the time, values of the variables are not Boolean, but they are the fitness values of the effects considered in the given environments, and those environments don't differ as in the example, but can range across a continuous scale depending on the values of a parameter; then the variable relevant for function ascription is the one that can be mapped onto this scale. Let's take a real example, taken from research on the function of sex. Sex is supposed to oppose Muller's ratchet (which means the eventual lethal accumulation of deleterious mutations in asexual organisms) (Maynard Smith 1979). Further, sex provides better genotypic variability when the environment is changing. Now, sex, in several actual changing environments, as well as genetic recombination, necessarily has those two effects because renewing the genotype at each zygote formation both prevents accumulation of deleterious mutations and makes zygotes more likely to match variations in the environment; those effects are less orthogonal than the polar bear's camouflage and heating effects (i.e. you can't have one without the other if you are in changing environments). Those two effects give a selective advantage to sexual individuals compared to asexual organisms, but the question is how to discriminate the function of sex, given that they go hand in hand, and so, in lots of known environments, they would have the same fitness. Now let's conceive of *several* environments, which differ regarding their variability. According to Hamilton, Axelrod and Tanese (1990), environmental changes provided by parasites are, given the timescale difference between life cycles of parasite and of hosts, of great amplitude. They simulated those environments. They showed that according to the degree or strength of parasites, hence of variability, sex is more or less favoured. This suggests that the defence against Muller's ratchet (which is independent of the degree of environmental variability conferred by parasites) is not relevant, and hence, the functional trait proper to sex is the providing of greater variability. (This study is also an exemplar in the sense that often counterfactual assumptions concerning environments are embodied in simulations.)

---

[9] This meaning of design is akin to that of Kitcher (1993). Buller (2002) also stresses the commitments to an idea of design proper to etiological theory analyses. Finally, in another framework, Wouters (2003), in a systematic analysis of functional explanation, describes what he calls design analysis. All those conceptions are distinct, and I don't undertake here a systematic comparison; see also Huneman (2007) on design as a necessary assumption of etiological theory.

My general point here is that provided that we assume an etiological theory of function, this theory cannot account for actual research if it does not consider how coarse-grained functional ascription (considering several effects as equally likely candidates for having been selected for) leads to fine-grained functional ascription, which disambiguates the bundle of effects through a specific method. In the above case, we have an implicit sophisticated counterfactual analysis. Let's write D for the Muller's ratchet avoiding effect and F for the variability-generating effect. The counterfactual here is about the sets of environments. Suppose we are in an environment E′, much more variable than the focal environment E that we are considering. Then, if D is the function of sex, sex would not be more likely to be selected in E′ than in E. If F is the function of sex, then sex is more likely selected in E′ than in E. So, finally, if the function of sex were D and not F, then sex would not have the same *increasing pattern of selection across a set of possible environments* as the one shown in the Hamilton et al. study. So F, the variability-generating effect, is the property enhancing fitness rather than D, the Muller's ratchet avoidance effect.

Notice first that this method mostly amounts to research on the maintenance of traits rather than on their origin (Reeve and Sherman 1993). Here, as in any research on maintenance, the trait is compared to possible variants rather than actual variants, whereas evolutionary research on origins is only concerned with variants that did actually occur. In the former case, no one can be sure that selection indeed acted at the origin of the trait and then explains why it came to the fore. In the toy case, it is perfectly possible that there had been no selection since there were no variants X and Y (see also Lewens 2004). This is the general problem with optimality investigations, since they do not rely on any historical evidence.[10]

In this regard, it is not surprising to see such method at work in the context of behavioural ecology. One striking case concerns the determination of the function of the silent bared-teeth display (SBTD) and the crest raise (CR) in the mandrills (Lairdre and Yozinski 2005). Here, we actually find a close variety of the counterfactual design method. Those behaviours have several possible correlated effects, which are plausible candidates for being their functions and which determine whether they are different signals or two grades of a same signal.[11] The authors "consider four possible functions: threat, submissive, conciliatory, and ambivalent".

[10] For example, L.C. Rome (1998) studies the construction of muscles in order to test the symmorphosis hypothesis. The idea that "in fact muscles are tightly matched to their function" is plausible since there is a "large disadvantage associated with using the wrong muscle type for a given activity". The experience reveals that optimal frequency of the power production by the muscle taken in isolation matches quite well with the actual frequency of the use of muscles. "Because of the large disadvantage associated with using the wrong muscle type for a given activity, it is likely that in fact muscles are tightly matched to their function. The tightness of this matching can be empirically determined by plotting optimal frequency of the isolated muscle power production versus the frequency at which the muscle is used in vivo". However, some more investigations are needed, he recognises, to confirm that in fact selection optimised the setting of muscles.

[11] I rather leave this second question aside in this case study.

The study parses the possible immediate environments in several kinds of interaction: "allogrooming, copulation, play, and agonism". Hence, those "four interactions (...) were used to generate ten mutually exclusive and exhaustive contexts: prior-to-groom (1 min prior to allogrooming); during-groom; prior-to-copulate (1 min prior to copulation); during-copulate; prior-to-play (1 min prior to play); during-play; prior-to-agonism (10 s prior to agonism); during-agonism (from agonism's start up to the last overtly aggressive or threat signal within an agonistic interaction); after-agonism (from the last overtly aggressive or threat signal within an agonistic interaction to 15 s after agonism's end); and nothing (a 30 min period in which none of the other nine contexts occurred)" (ibid., 146). The question is then to determine in which contexts those behaviours are most likely to be triggered.

Each candidate function yields some predictions about whether the signal (SBTD) is more or less likely to appear in one context than in another. For instance for a conciliatory signal, our predictions are as follow:

1. More likely to occur in the prior-to-groom vs. prior-to-agonism context
2. More likely to occur in the during-groom vs. during-agonism context
3. More likely to occur in the prior-to-copulate vs. prior-to-agonism context
4. More likely to occur in the during-copulate vs. during-agonism context
5. More likely to occur in the prior-to-play vs. prior-to-agonism context
6. More likely to occur in the during-play vs. during-agonism context
7. More likely to occur in the after-agonism vs. prior-to-agonism context. (ibid., 148)

The biologists therefore analysed interactions in all of those contexts; the final statistical pattern mostly matched the above prediction, which means that the function of the signal is the conciliatory one.[12]

Here again, put abstractly, the reasoning is the following: all the candidate functions (A1...A4) are logically intertwined, but if the behaviour had effects A2, A3 and A4 but not A1, then there would not be such a statistical pattern of occurrences (exactly like: if sex were not selected for the effect of generating variability, the pattern of selection of sexual reproduction across diversely variable environments would not be the same). So in the context of maintenance questions proper to behavioural ecology, the counterfactual design method is instantiated in various ways in order to disambiguate the logically related candidate functions of some behaviour. Moreover, functional *explanations* are relying on this method: suppose that I wanted to explain how mandrills deal with crisis: I would then resort to the previous analysis and cite the SBTD behaviour as part of a general strategy (which would involve CR specifically, but several sequences of richer behaviours).

For sure, Millikan (1989) said that to be a function is not a probabilistic or a dispositional statement; it is a statement about history. Yet from the perspective of the counterfactual design method here described, even though we can discriminate between several candidates to functional ascription, there is no history here because the historical role of selection is not established, and the ascription is compatible

[12] The fact that the statistical pattern of CR is not so different, but has less significance, leads the author to consider it a graded form of the SBTD.

with a wide variety of possible histories of origin. In this sense, it seems that this method contradicts the etiological theory. However, to the extent that part of a functional discourse in biology uses those counterfactual design arguments, often in the form of simulations, we cannot throw them out of our theory of function if we don't view such theory (unlike Millikan's) as a stipulating definition. So here, as in the case of the functional organisation schema, we supplement a functional ascription along the lines of SE theory by an approach that is not in the scope of this theory. The etiological theory alone cannot account for the entire functional ascription in its explanatory context.

3.3 The Comparative Method

Comparison between different and more or less distant species is an overwhelmingly common method in biology (e.g. Harvey and Pagel 1991). Concerning functional ascriptions, its relevance first concerns the very identification of a trait as having a function. Suppose that the same trait arises in two different species, for example, species from two clades. This makes it very unlikely that the same phylogenetic constraint or process of drift could be at its origin, so it must have been the result of the same selective pressure acting in the same way upon two phylogenetically distinct populations in two perhaps distinct environments. Eyes are a typical example of this: they have evolved more than 20 times in evolution.

Now, if we turn to our problems, suppose that, apart from our trait T in species S that has two correlated effects of same fitness Y and X, there exists species S′ and S″ in other clades that bear the same trait T. Not only does the comparison provide evidence for the fact that T has a SE function, but it helps to solve the bundle of effects problem. Indeed, suppose that T is a detector. Given that species S, S′ and S″ are very different, it is likely that their environments are a bit different. Suppose that in S, T detects Y that is a; then if in species S′, T detects Y which is a′; and in species S″, T detects Y which is a″, we would say that T in S evolved because of its property of detecting Y rather than because of its detecting a (see Table 1.) This is implied by the fact that, due to the convergence, we consider that X is here because of the same selective pressure in the three species, while the *a*s differ among those species.

Compared to the counterfactual design method, this method yields ascription of "stimulus", as Neander says, rather than benefit, as the genuine function of the trait. To show it, think of various animals, with devices that detect moving dots (the Y's in

Table 1 Example of comparative method: different individuals in three distant species S, S′, S″ have the same detecting trait T detecting Y, but it detects various correlated things A, A′, A″ in their environments

Species	S	S′	S″
First target of detection	Y	Y	Y
Second target of detection	A	A′	A″

the table), those dots being the various preys in their environments (the a's in the table). This is precisely Neander's position; notice however that, contrary to what she argued, this point doesn't concern all functional ascriptions but only those that are embedded in the comparative method sketched here.

Now, even if detectors with their effect-stimulus and effect-benefit are an easy toy case, this method is much more pervasive and concerns even cases where traits are not so vertically connected. In fact, each time you have rival hypotheses concerning two effects likely to be "the function" of something because they are connected so that they contribute equally to fitness, considering the same traits in distinct species allows one to distinguish these effects since they won't be connected anymore. Hence, this method enables us to identify the function of a trait in a given species. For example, it has been used to discriminate hypotheses about sexual dimorphism concerning body size in primates. The rival hypotheses were increasing chances for sexual selection (Darwin 1871) and enabling a separation of niches for exploiting resources (Selander 1966). In some species, sexual dimorphism did both things, which seemed to contribute in the same way to fitness, so according to the etiological theory, the function of dimorphism could be both. But when you compare several species of primates, it appears that highly sexually linked body size difference is mostly found in species where the mating system is polygamy; hence, the function of body size is supposed to "attract mates", as hypothesised by Darwin's initial theory.

To this extent, appeal to the comparative method allows biologists to *confirm hypotheses on functional ascription.* Laughlin (1998) considers the function of potassium channels in eyes. "The potassium channels have precisely the combination of properties required to match the gain and response speed of the membrane to the photo transduction cascades: a high gain and slow response in the dark, and a low gain and fast response when depolarized by light". That leads to the conclusion that "these potassium channels appear to have been selected for this regulatory role". However, this does not buffer the hypothesis against rival hypotheses that would consider potassium channels to have been selected for an effect P regularly connected to its role of matching gain and response speed.

Here enters the comparative method: "*Comparative studies*, a useful tool for *probing design*, show that slowly flying *Diptera* have photoreceptors that fail to speed up with light adaptation. In the absence of fast-moving signals, this slow response is better, and the photoreceptors are using inactivating potassium channels to save energy, even in bright light" (my emphasis).

Let's unpack the reasoning implicit here. I write M for the modulation between gain in photo transduction and speed of response and P for some other property correlated to this modulation. Organisms of a slower parent species do not have to cope with the need to modulate signals because if they fly slowly, the signals around them are slow. If M is selected, then in such a species, the effect of speeding up transmission with increases in light intensity should *not* occur, and therefore, we would not expect a fast response in bright light. If P is selected, then given that those variations in light are not relevant to the selective advantages of P, in such parent species, we should still see the modulation M of speed of transmission in bright light correlated to P since M is there because of the fact that P has been selected. The former case happens because even in bright light, potassium channels are used to maximise the gain of energy by slowly flying species, so it is plausible to think that the selected property is indeed the modulation M and not one of its correlates.

This example is also interesting since it contradicts an approach of the comparative method that would easily arise when one considers it in relation to the previous one. I said that counterfactual design method mostly concerns the maintenance of traits; the comparative method as I presented it seems most suited to the birth of traits in a clade. However, in the example of the eyes and their potassium channels, everything is compatible with the hypothesis that those potassium channels appeared once in the eye of some ancient species and then got several different functions in different species. This goes against the interpretation of the comparative method as oriented only towards discovering genealogies of traits (and not maintenance).

A clearer example is given by E. De Margerie (2002, 2006) and Margerie et al. (2005) who considered the shape of the bones in birds' wings as a trait. Traditionally their being hollow was thought to have the function of enabling the bird to fly. But Swarts et al. (1992) suggested that their bony structure might rather be an adaptation to torsion generated upon the birds' wings by the flight. De Margerie tested this hypothesis by a comparative approach. Optimal histological values of bony tissue are not the same if the bone is adapted to *torsion* or if it's adapted to *flexion*, as it is the case when one thinks that its hollowness facilitates flight. So we have two conceptions: hollowness is selected because it allows flight by resisting flexion, or it is selected because it facilitates flight by allowing torsion. No one doubts that the function of wings is to fly, but the question is the function of the *shape of the bones*—why do they facilitate flying? It happens in many species that bones both resist flexion and torsion, so if a biologist considers the function of the hollowness of bones in such a species, she faces our case of two equally fit connected effects of one trait, preventing a simple ascription along the lines of the etiological theory. However, if you compare several species of birds according to the value of some parameters, measuring the torsion resistance on several bones (2006, fig. 5, 626), you see not only that torsion resistance is dominant in those bones most exposed to torsion (ulna, humerus) rather than in the others but also that in some species torsion resistance is weaker. Yet those species precisely (*Diomedea melanophris*, *Macronectes giganteus*, *Procellaria aequinoctialis*) are less exposed to the torsion of their wings' bones because of their way of flying (namely, gliding), and the lengthening of their wings makes them more subject to flexion. Therefore, preventing torsion seems "one of the strongest selective pressures on the skeletal adaptation to fly by vertebrates" (ibid., 627).

Here then, we see that the comparative method helps to distinguish several effects that in one species could, on the sole basis of the etiological theory, count as equally good candidates to be the function of a trait (namely, torsion resistance and flight). It achieves this result by construing an explanation that has to be distinguished from the two previous ones: here, the explanation doesn't aim at understanding the presence of a trait in an organism or the designedness character of this trait in relation to the organism but rather the frequency of a trait in several related or unrelated species, which ultimately is a sort of partial taxonomic pattern. This concern with actual phylogenetic and taxonomic order justifies the main difference from the counterfactual design method, namely, that the latter, being an optimality strategy, uses possible variants, while the comparative method uses actual variants (yet its result depends upon the definition of the class of the comparison, which is the general worry of this method).

3.4 Confronting Methods

So in the end, we have several ways to disambiguate the cluster of properties likely to be the function of the trait T, but those ways will not lead to the same determination of the function of T. This is most easily shown by the toy case of the detector because the counterfactual design approach ascribes the benefit target as a function, contrary to the comparative method. The attempts of Dretske (1986), Neander (1995), Agar (1993), Price (1998), etc. to solve the so-called determination of content problem seem to assume that, properly understood, the determination of the function by the etiological theory will provide *one direct way to determine what the function of T is*. For example, Agar (1993) rightly emphasises that appeal to counterfactuals disambiguates rival candidate hypotheses for a functional ascription—yet he failed to see that this is not the only method through which the function can be revealed. The general mistake in those attempts is that there is not such a directly available way, given that if one wants to discriminate within the cluster of effects, one must supplement the etiological ascription of function by one of those three methods, the choice of which is not provided by the theory itself but, on the contrary, relies on one's explanatory interests concerning the functional trait debated. Clearly, if the connection between the candidate effects is contingent in the sense of world-dependent, meaning that in our world, all occurrences of each one are connected but that it could be otherwise in another possible world, then the only appropriate method is the counterfactual one. However, in all other cases of less metaphysical context-dependence, the choice is not constrained and thereby relies on *which explanations* one is willing to undertake with her identifications of functions, be it the revealing of a general design of an organism, the unravelling of the designedness or optimality of its design or finally the establishment of a taxonomic pattern of functional traits among clades or species.

Now, what are the consequences of the fact that functional ascriptions under the three methods can diverge, when it comes to a selected-effects theory of functions? My point was that the various methodologies of functional explanations must be taken into account if one wants to solve the bundle of effects problem, which prevents the etiological theory of function *alone* from making sense of fine-grained functional ascriptions. This raises two questions: Are those methods actually relevant for establishing selection history? And if so, is it legitimate to include them in a selected-effects account of functions (or do they undermine several of the requisites of the SE theory)?

3.4.1 Divergent Results and Selection

First, about selection history, it is unclear whether each of the methods for functional explanation establishes a selection history. Especially in the case of the mandrills and their facial expression, the counterfactual design method does not reveal the selective origins of the signal; it may be that it arose because of one of the various candidate functions, but the fact is that the actual function determined by this method is the one which would plausibly cause its maintenance. To this extent, not all the methods here are such that they provide an access to a fine-grained understanding of selective *history*. On the contrary, often they *don't* concern the selective history (in the sense of the history of origins) so that a plausible consequence of taking the methods of functional explanation into account would be that one should prefer a "modern history" view of etiology, *sensu* (Godfrey-Smith 1993), if the etiological view of function is still to be held.

Many more methods to infer selection are examined in Endler (1986), but the aim of this chapter concerns only the bundle of effects problem in ascribing functions, not the knowledge of selection in general. Hence, considering three methods, although not exhaustively, was enough to raise the problem for the etiological theory and now to revise it.

As to the first method, the design one, it aims at supplementing the functional ascription (*sensu* the etiological view) when it's too coarse grained, but it does not say anything about selection, so it disambiguates the bundle of effects problem, but it does not help distinguish between correlated candidate traits for being selected-for. Including it *within* an account of functional ascription therefore raises a real issue for the etiological view, since it does not remain *only* etiological.

However, what about a possible divergence between methods? It is indeed not a systematic one. In some cases, the three methods would yield the same functional ascription. In some other cases, not all methods are even available; for example, in the mandrill case, I doubt the comparative method would make sense. So, apart from the fact that organism-design method is not directly informative about selection, there is no reasonable doubt that the other methods considered say something about what has been selected, what is under selection and why. When the divergence occurs, it may be the case that there is no principled way to decide "for what" there has been (or is) selection, a conclusion akin to Lewens' (2009) sophisticated analysis of the difference between selection-for and selection as a force (measured in population genetics).

Speaking very generally about establishing the facts of selection, it seems that there is a continuum of possible cases. First, there are cases where we know that there has been selection, but we don't know what for. The clearest example is the detection of the signature of selection in the genome. The genetic patterns of variation caused by selection and by drift are different, and we have tests such as the Kreitman test to detect the one due to selection. But it does not require that we know what the genetic sequences are coding for at all, so when we detect the signature of selection, most of the time we know that there has been selection but we don't know what there has been selection-for. The other extreme pole of the continuum is when we know that there has been selection, and we can know what there has been selection-for—metal tolerance in plants is one complete example, detailed by Brandon (1996). In between, there are all cases where we know that there has been selection, but we cannot disambiguate several possible candidates in principle, at least without already having an explanatory interest or a guiding question in mind. This is what happens with the cases presented above where the available methods diverge in the functional ascriptions they yield.

Nowadays, such considerations echo a hot debate raised by Fodor's paper (2008) against Darwinism.[13] Fodor's point specifically concerns the bundle of effects problem; he argues that there is no principled way to say what there has been selection-for, between a trait and another coextensional to it. (Fodor uses the "selection-for/selection-of" distinction first articulated by Sober (1984)); in a word, he says that we can never know what there has been selection-for, even if we knew there has been selection-of.) Given that I've considered some methods to solve the bundle of effects problem and that those methods indeed distinguish between a selected trait and its correlate in a counterfactual or a comparative manner, in many cases, we are in the middle of the continuum discussed here. So we have some knowledge of what there has been selection-for; against Fodor, there is *no a priori reason* to say that we can't know about the facts of selection.

Fodor and Piattelli-Palmarini (2010) have attracted so many replies (e.g. Sober 2008, 2010; Okasha 2010; Lewens 2008; Godfrey-Smith 2010) that basically all major answers to be made about both general issues in philosophy of science (laws and counterfactuals) and evolutionary biology have been formulated. Given that my paper touches a parallel problem with the bundle of effects issue for functional ascriptions, I mention some of its consequences for Fodor's claim in passing. Fodor has one main argument, which is to say that selection statements are intentional and not extensional contexts and that for this reason, they can't be considered as correct or at least unproblematic causal statements, whereas pinpointing selection-for should be a causal statement. The only way to overcome the problem raised by intentional contexts is that there should be laws of selection-for, but there are not (because of the context sensitivity of selection[14]) and then we cannot write the counterfactual-supporting statements that in other causal contexts would allow us to state causal ascriptions. But the methodological considerations shown here, together with my examples, support the view that, indeed, there is no particular problem with using counterfactual reasoning about selection, so no reason to a priori reject the project of asking for what there is selection.[15]

[13] The bulk of this paper was written in 2005–2006, before Fodor's papers were published. Under the suggestion of an anonymous referee who pointed out that the issues raised here are also debated in the discussion following Fodor and Piattelli-Palmarini's book publication, I situate here my views relative to Fodor's.

[14] Among other quotes, it's "quite likely there aren't laws of selection. That's because who wins a t1 v. t2 competition is massively context sensitive". (Fodor 2008)

[15] See Sober's (2010) specific answer to the point about laws.

Most generally, Fodor's argument criticises selection-for ascriptions as causal statements. He contrasts them with uncontroversial causal statements, which are such that we can easily distinguish two correlated properties, thanks to a simple counterfactual test. The example he gives is the scotch and ice drink—which made me drunk: even if ice and scotch were involved in the cause of my sickness, one can easily distinguish them by testing what it would be with whisky and not ice, and vice versa. However, this is not the whole story. "Being whisky" entails being "alcoholic beverage at 40%", "beverage with more than 35% of alcohol", etc. So you have many plausible candidate causal properties nested—and not only two. Fodor's claim that the whisky-on-the-rocks case allows simple causal statements means that this state of things is very different to the case of selection-for and our bundle of effects issue (with contingently or necessary correlated properties). But if you consider such nesting of properties, it's not so different. The counterfactual tests needed to handle the whisky case are in fact not so easy to design—for example, it might be that whisky is such that I will be more sensitive to a specific amount of alcohol than if alcohol were included in another liquor. So whisky, *as whisky* could certainly be causally relevant. But still, it has to be tested against possible worlds where *other specified drinks* are considered, and then other counterfactual tests have to be made up. Finally, I just want to point out that causal statements in general, contrary to Fodor's claim, are not so clearly extensional and that, on the contrary, selection-for does not in principle face more critical issues. Too stringent requirements on laws and explanations, as Fodor deploys in his first argument against selection-for, would surely also undermine many causal claims outside of the scope of natural selection, but few people will care about specifying the exact causal powers of whisky as compared to those of "Scottish whisky" or "strong alcohol" (or, more exactly, some would if they are biochemists or doing studies about genetics, epidemiology of alcoholism and cultures of drinking; here also, explanatory interests matter). What is important, actually, is that evolutionary biologists elaborated several methods to distinguish correlated properties according to their explanatory interests and that functional explanations are developed along those lines.

Finally, the facts of selection can be more or less coarse-grained. Establishing coarse-grained facts of selection involves robust reasoning and models, especially in population genetics, that is, when people consider alleles or genotypes, leaving aside the ecological reasons why they have the fitness values that they have, and model their dynamics in mathematical terms. Sober (2010) is wholly right to highlight that Fodor leaves out the population genetics model, where general causal facts about selection are constantly established. In this sense, there is often no problem in science with selection-for. Only a metaphysician would object to a biologist who says we know that the "frog's perception devices were selected for their ability to catch moving flies" (without being willing to separate those). But when it comes to ascribing functions and, especially, to make philosophical sense of what functions are and whether they are part of the furniture of the world or not, then the bundle of effects problem will challenge the theoretician.

3.4.2 Etiological Theory?

Facts of selection are in the world. The issue we are facing now is to decide whether a realist theory like the etiological one can use those facts of selection to claim that there are unambiguous facts of the matter which yield all functional ascription, given that, as I have shown, functional ascriptions do not stem univocally from fitness.

To sum up, if there is to be one real function of a trait T, it's not enough to define the concept of function according to the etiological theory. We also have to choose one approach for fine-graining the function of a trait; this choice is not prescribed by the etiological definition of the function, but the final attribution of the function of T will depend on it.

The immediate consequence of such analysis is that either the *discriminative* or the *realist* requirement of the etiological theory seems too strong: either the theory is not discriminative enough, or in order to establish *the function* of an item, a methodological strategy has to be specified independently of the etiological theory. Although the function of the item, being based on its causal history, is not likely to be determined by us (unlike in the case of systemic theories where functions are internal to a choice of an explanatory target from the start), such function will somehow depend on this strategy because the three different methodologies available and examined here may yield different results.

So the ambition of justifying that there actually *are* functions in nature—ambition that etiological theorists frequently opposed to systemic theories—is not entirely fulfilled: those functions are never completely free of the explanatory interest that at some point casts a light on them. Of course, when all three available methods yield the same functional ascription, it's plausible to hold a realist view about functions here, but not all biological functions can be analysed in this way if one considers cases where the various available methods diverge.

The last problem is that it's not clear whether the etiological theory is still an etiological one if we consider that the ways of solving the bundle of effects problem do not specify a proper etiology for the trait, as suggested above. If, for example, one resorts to the design method, what grounds the functional ascription is not the selection process, but something more than that. It may not be correct to think that fitness grounds a coarse-grained functional ascription and that some method for functional explanation makes it fine grained, even if I presented it in this manner in the beginning. Here the problem is that the final fine-grained ascription of function is not any more defined by selection only—especially, not necessarily by the selective *history* of the trait, since it's often considerations about its maintenance (in the counterfactual and often the comparative methods) that allow one to specify the function.

To this extent, the etiological theory cannot properly be said to be a selected-effects theory of function because selection is not enough to ascribe functions (once again, from the viewpoint of evolutionary biology itself—as opposed to the conceptual analysis of functional concepts—the question of correlated, coextensive, co-selected properties is not a real issue since coarse-grained definition of properties is acceptable). The only general characterisation of etiological theory of function left here is that considerations about selection—especially maintenance selection—and fitness are necessary for functional ascriptions. From this point on, let's examine what are the characteristics of any etiological theory of functions.

Among the three main requisites proper to SE (leaving *discriminative* untouched), the *realism* requisite has surely to be weakened. The actual functional ascription relies on some explanatory considerations underpinning the choice of the method that will yield the fine-grained functional ascriptions and solve the bundle of effects problem. Even if there are facts of selection, which justifies some realism for the etiological theory of function, the fine-grained specification of functions is not wholly realist; it requires taking into account our explanatory interests. This is not the "weak theory" in the sense of Buller (1998) because what is weakened is the realist requisite, not the focus on selection.[16] In any case, "the function of T", when we suppose that there is one genuine function, cannot be independent of the choice of explanatory strategy, even if it is not the case that any function is likely to be ascribed and any strategy to be appropriate.

Now, what about the two other requisites, *explanatory* and *normative*? As to *explanatory*, it seems that nothing is changed: in any case, those functional ascriptions, even when supplemented by one of the methods discussed here, are explanatory regarding the functional item. But *normative* may not be so immune to the considerations exposed here. The reason why functions are a normative concept, according to the etiological theory, is indeed tied to its realist stance: because the function Y of X is given by its selective history, which is a real fact, then tokens X that are not doing what type X objects have been selected for are not doing what their function is (as tokens of X). But if what appears to be the function of X is determined not only by the facts of selection but also by some explanatory choices, then it's less obvious that Y is a norm for all tokens X. So, even if the main consequence of such analysis is that the *realism* of the etiological theory has to be deflated, it may be that the *normative* requisite has also to be reconsidered.

4 Conclusion

In the last section, I have shown that if there is to be one real function of a trait T, it's not enough to define the concept of function according to the etiological theory. We also have to choose one approach for identifying the function of a trait at a fine grain; this choice is not prescribed by the etiological definition of the function, but the final attribution of the function of T will depend on it.

The final result of such an investigation into functional explanations in evolutionary biology, their various methods and the consequences upon the bundle of

[16] Buller's weak theory connects functional ascription to evolutionary history in general, not only selective history.

effects problem is that, contrary to its ambitions, the etiological theory cannot fulfil all its requisites at the same time. If it wants to account for fine-grained functional ascriptions in biology and then keep the ideal of being *discriminative*, it has to weaken its *realist* ambitions. This clearly follows on from the fact that functional explanations, under some given explanatory framework, have to be considered in order to specify a precise and discriminatory functional ascription.

From this viewpoint, the famous gap between etiological theories and systemic theories of functions *à la* Cummins is less huge than expected; none of them can avoid reference to an explanatory interest underpinning functional ascriptions, even if this is on the forefront of the sole systemic theory.

Also, the *normative* requisite, which is supposed to be a clear sign of this gap between both theories, is not absolutely fulfilled by the etiological theory; in this sense, it might be that, provided that one could also sketch a possible account of normativity in a systemic theory (which would of course not be plainly realist and naturalist), both views of function could also be articulated within a single project of accounting for the normativity claims embedded in functional ascriptions.

Finally, the very name of the etiological theory exhibits a reference to selection in general as a necessary reference for functional ascriptions, but not as a sufficient one, so that it is much more controversial to call it selected-effects theory of functions, and the notion of etiology is itself not to be taken at face value because, as we have seen, disambiguating functional ascriptions require taking into account important nonhistorical facts and structures.[17]

References

Agar, N. 1993. What do frogs really believe? *Australasian Journal of Philosophy* 71: 1–12.

Ariew, A. 2003. Ernst Mayr's 'ultimate/proximate' distinction reconsidered and reconstructed. *Biology and Philosophy* 18(4): 553–565.

Beatty, J. 1994. The proximate/ultimate distinction in the multiple careers of Ernst Mayr. *Biology and Philosophy* 9: 333–356.

Brandon, R. 1996. *Adaptation and environment*. New York: Oxford University Press.

Buller, D. 1998. Etiological theories of function: A geographical survey. *Biology and Philosophy* 13: 505–527.

Buller, D. 1999. Natural teleology. In *Function, selection, and design*, ed. D. Buller, 1–27. Albany: SUNY Press.

Buller, D. 2002. Function and design revisited. In *Functions: New essays in the philosophy of psychology and biology*, ed. A. Ariew, R. Cummins, and M. Perlman, 222–243. New York: Oxford University Press.

Cummins, R. 1975. Functional analysis. *Journal of Philosophy* 72: 741–765.

Darwin, C. 1871. *The descent of man, and selection in relation to sex*. London: John Murray.

[17] The first version of this chapter has been presented at a symposium on functional explanations at the ISHPSSB meeting in Guelph 2005; I thank the audience for the insightful questions there. I also thank Françoise Longy for her careful reading and suggestions and Marshall Abrams who also did language checking.

De Margerie, E. 2002. Laminar bone as an adaptation to torsional loads in flapping flight. *Journal of Anatomy* 201: 521–526.
De Margerie, E. 2006. Fonction biomécanique des microstructures osseuses chez le oiseaux. *Comptes rendus de l'académie des sciences Palévol* 5(3–4): 619–628.
De Margerie, E., S. Sanchez, J. Cubo, and J. Castanet. 2005. Torsional resistance as a principal component of the structural design of long bones: comparative multivariate analysis in birds. *The Anatomical Record. Part A, Discoveries in Molecular, Cellular, and Evolutionary Biology* 282A: 49–66.
Dretske, F. 1986. Misrepresentation. In *Belief, form, content and function*, ed. R. Bogdan. Oxford: Clarendon.
Enç, B. 2002. Indeterminacy of function attributions. In *Functions*, ed. A. Ariew, R. Cummins, and M. Perlman, 291–313. Oxford: Oxford University Press.
Endler, J. 1986. *Natural selection in the wild*. Princeton: Princeton University Press.
Fodor, J. 2008. Against Darwinism. *Mind and Language* 23: 1–24.
Fodor, J., and M. Piattelli-Palmarini. 2010. *What Darwin got wrong*. New York: Farrar, Straus, & Giroux.
Godfrey-Smith, P. 1993. Functions: Consensus without unity. *Pacific Philosophical Quarterly* 74: 196–208.
Godfrey-Smith, P. 1994. A modern history theory of functions. *Nous* 28: 344–362.
Godfrey-Smith, P. 2010. Review of what Darwin got wrong, by Jerry Fodor and Massimo Piattelli-Palmarini. *London Review of Books* 32(13): 29–30.
Hamilton, W., R. Axelrod, and R. Tanese. 1990. Sexual reproduction as an adaptation to resist parasites (a review). *Proceedings of the National Academy of Sciences USA* 87(9): 3566–3573.
Harvey, P.H., and M.D. Pagel. 1991. *The comparative method in evolutionary biology*. Oxford: Oxford University Press.
Horan, B. 1989. Functional explanations in sociobiology. *Biology and Philosophy* 4(2): 131–158.
Huneman, P. 2007. Pourquoi ne fait-on pas de montres en caoutchouc. Limites de la détermination fonctionnelle des parties d'un tout. In *Le tout et les parties*, ed. T. Martin, 75–87. Paris: Vuibert.
Keynes, R. 1998. The design of peripheral nerves fibers. In *Principles of animal design. The optimization and symmorphosis debate*, ed. E. Weibel, R. Taylor, and L. Bolis, 271–277. Cambridge: Cambridge University Press.
Kitcher, P. 1993. Function and design. *Midwest Studies in Philosophy* 18(1): 379–397.
Lairdre, M., and J. Yozinski. 2005. The silent bared-teeth face and the crest-raise of the mandrill (*Mandrillus sphinx*): A contextual analysis of signal function. *Ethology* 111: 143–157.
Laughlin, S. 1998. Observing design with compound eyes. In *Principles of animal design. The optimization and symmorphosis debate*, ed. E. Weibel, R. Taylor, and L. Bolis, 278–287. Cambridge: Cambridge University Press.
Lewens, T. 2004. *Organisms and artifacts: Design in nature and elsewhere*. Cambridge, MA: MIT Press.
Lewens, T. 2008. Reply to Jerry Fodor "Why don't pigs have wings". *London Review of Book* 27: 1.
Lewens, T. 2009. The natures of selection. *The British Journal for the Philosophy of Science* 61(2): 1–21.
Mayr, E. 1961. Cause and effect in biology. *Science* 134: 1501–1506.
Maynard, Smith J. 1979. *The evolution of sex*. Cambridge: Cambridge University Press.
Millikan, R.G. 1984. *Language, thought, and other biological categories: New foundations for realism*. Cambridge, MA: MIT Press.
Millikan, R.G. 1989. Biosemantics. *The Journal of Philosophy* 86(6): 28.
Nagel, E. 1961. *The structure of science*. London: Routledge & Kegan Paul.
Neander, K. 1990. Dretske's innate modesty. *Australasian Journal of Philosophy* 74(2): 258–74.
Neander, K. 1991a. The teleological notion of function. *Australasian Journal of Philosophy* 69: 454–468.
Neander, K. 1991b. Functions as selected effects: The conceptual analyst's defense. *Philosophy of Science* 58: 168–184.
Neander, K. 1995. Misrepresenting and malfunctioning. *Philosophical Studies* 79: 111.

Nesse, R., and G.C. Williams. 1995. Why We Get Sick: The New Science of Darwinian Medicine. London: Vintage.

Okasha, S. 2010. Review of what Darwin got wrong, by Jerry Fodor and Massimo Piattelli-Palmarini. *Times Literary Supplement*, March 26.

Price, C. 1998. Determinate functions. *Nous* 32: 54–75.

Profet, M. 1992. Pregnancy sickness as adaptation: A deterrent to maternal ingestion of teratogens. In *The adapted mind: Evolutionary psychology and the generation of culture*, ed. J. Barrow, L. Cosmides, and J. Tooby, 327–365. Oxford: Oxford University Press.

Reeve, H.K., and P.W. Sherman. 1993. *Adaptation* and the goals of evolutionary research. *Quarterly Review of Biology* 68: 1–32.

Rome, L.C. 1998. Matching muscle performance to changing demand. In *Principles of animal design. The optimization and symmorphosis debate*, ed. E. Weibel, R. Taylor, and L. Bolis, 103–113. Cambridge: Cambridge University Press.

Selander, R.K. 1966. Sexual dimorphism and differential niche utilization in birds. *Condor* 68: 113–51.

Sober, E. 1984. *The nature of selection*. Cambridge, MA: MIT Press.

Sober, E. 2008. Fodor's *Bubbe Meise* against Darwinism. *Mind and Language* 23: 42–49.

Sober, E. 2010. Natural selection, causality, and laws: What Fodor and Piattelli-Palmarini got wrong. *Philosophy of Science* 77: 594–607.

Stearns, S. 1992. *The evolution of life histories*. Oxford: Oxford University Press.

Swarts, S., M.B. Bennet, and D.R. Carrier. 1992. Wing bones stresses in free flying bats and the evolution of skeletal design for flight. *Nature* 359: 726–729.

Vermaas, P.E., and W. Houkes. 2003. Ascribing functions to technical artefacts: A challenge to etiological accounts of functions. *British Journal for the Philosophy of Science* 54: 261–289.

Walsh, D. 1996. Fitness and function. *British Journal for the Philosophy of Science* 47: 553–574.

Wimsatt, W. 2002. Functional organisation, analogy and inference. In *Functions*, ed. A. Ariew, R. Cummins, and M. Perlman, 173–221. Oxford: Oxford University Press.

Wouters, A.G. 2003. Four notions of biological function. *Studies in History and Philosophy of Biology and Biomedical Science* 34(4): 633–668. doi:10.1016/j.shpsc.2003.09.006.

Wright, L. 1973. Functions. *The Philosophical Review* 82: 139–168.

Part III
Psychology, Philosophy of Mind and Technology: Functions in a Man's World – Metaphysics, Function and Philosophy of Mind

Functions and Mechanisms: A Perspectivalist View

Carl F. Craver

Abstract Though the mechanical philosophy is traditionally associated with the rejection of teleological description and explanation, the theories of the contemporary physiological sciences, such as neuroscience, are replete with both functional and mechanistic descriptions. I explore the relationship between these two stances, showing how functional description contributes to the search for mechanisms. I discuss three ways that functional descriptions contribute to the explanations and mechanistic theories in contemporary neuroscience: as a way of tersely indicating an etiological explanation, as a way of framing constitutive explanations, and as a way of explaining the item by situating it within higher-level mechanisms. This account of functional description is ineliminably perspectival in the sense that it relies ultimately on decisions by an observer about what matters or is of interest in the system they study.

1 Introduction

In its most austere and demanding forms, the mechanical philosophy insists on a disenchanted world explicable without remainder in terms of basic causal principles. Though mechanical philosophers differ from one another about which causal principles are basic (size, shape, and motion for Descartes; attraction and repulsion

Thanks to Lindley Darden, Marie Kaiser, Sarah Malanowski, and an anonymous referee for comments on earlier drafts; Pamela Speh for graphic design; and Tamara Casanova, Mindy Danner, and Kimberly Mount for administrative support. This project was finalized during a stay in spring 2011 with the DFG Forschergruppe on Causation and Explanation at the Universität zu Köln.

C.F. Craver (✉)
Philosophy-Neuroscience-Psychology Program,
Washington University, St. Louis, MO, USA
e-mail: ccraver@wustl.edu

P. Huneman (ed.), *Functions: selection and mechanisms*, Synthese Library,
Studies in Epistemology, Logic, Methodology, and Philosophy of Science 363,
DOI 10.1007/978-94-007-5304-4_8,

for du Bois-Reymond; conservation of energy for Helmholz), they univocally reject explanations that appeal to vital forces and final causes. Austere views such as these are commonly associated with the idea that the mechanical world is an aimless machine, churning blindly, without its own end or purpose, and also with the apparent historical opposition between functions and mechanisms as conceptual tools for understanding the natural world.[1]

Contrast this historical opposition of function and mechanism with the state of play in early twenty-first-century physiological sciences, such as neuroscience. In such fields, the language of mechanism is literally ubiquitous, and most scientists continue to demand that adequate explanations reveal the hidden mechanisms by which things work (Bechtel and Richardson 1993; Craver 2002, 2007; Craver and Darden 2001; Machamer et al. 2000). Yet the mechanical philosophy embodied in the explanatory practices of the twenty-first-century physiological sciences embraces functional descriptions as well. Consider some recent titles:

MicroRNAs: Genomics, biogenesis, mechanism, and function (Bartel 2004)
Mechanism and function of formins in the control of actin assembly (Good and Eck 2007)
Serpin structure, mechanism, and function (Gettins 2002)
Mechanisms and functional implications of adult neurogenesis (Zhao et al. 2008)

Theoretical terms such as vesicle, neurotransmitter, receptor, channel, and ocular dominance column are conspicuously functional, describing entities not in terms of size, shape, and motion but in terms of their job or role in the behavior of a system. This intermingling of functional and mechanistic descriptions is not limited to molecular and cellular phenomena. The doctrine of localization of function, a cornerstone of contemporary neuroscience, claims that discrete brain regions and brain mechanisms perform distinct functions.

This happy coexistence of functional and mechanistic descriptions in our contemporary physiological sciences suggests that the concept of mechanism, the concept of function, or both are significantly different from the way they were understood within the mechanical philosophy of the seventeenth century. In this chapter, I embrace a form of perspectivalism about both functions and mechanisms, one result of which is to narrow this historical gap. Just a early advocates of the mechanical philosophy insisted, I claim that the causal structure of the world is disenchanted and purposeless. Mechanistic and functional descriptions, in contrast, presuppose a vantage point on the causal structure of the world, a stance taken by intentional creatures when they single out certain preferred behaviors as worthy of explanation. Specifically, talk of functions and final causes is not legitimized by or reduced to privileged kinds of etiological histories (though some functions have

[1] Westfall describes the world of the mechanical philosophy as a "lifeless field knowing only brute blows of inert chunks of matter" (1973, 31; see also Westfall 1971; Shapin 1996). Historians have suggested that the opposition of mechanism and Aristotelian explanations in terms of forms and final causes oversimplifies the diversity of perspectives one finds in the seventeenth century and beyond (see especially Allen 2005; Des Chene 2005; Osler 2001).

such histories) or to certain special effects of the item in question. Rather, they are imposed from without by creatures seeking to understand how a given phenomenon of interest is situated in the causal structure of the world.

The philosophical project surrounding functions and mechanisms so conceived is not to find a way of building them into the causal bedrock of the world but of understanding the essential role that these notions play in physiological sciences such as neuroscience. One project is to understand how functional and mechanistic descriptions are related to one another in physiological sciences. In particular, I stress the roles that functional description plays in the effort to construct multilevel mechanistic theories (Craver 2002). A second project is to show how functional descriptions can be explanatory even when there is no etiological story to tell about how the functional item came to be. I argue that functional description can serve as a form of causal-mechanical explanation; it is a means of situating an item in the causal structure of the world (Salmon 1984). A third project is to make explicit the evidential criteria by which functional and mechanistic descriptions are evaluated. I argue that functional attributions are contentful to the extent that they can be cashed out in a detailed description of how an item is organized into a higher-level mechanism and that one has good evidence for one's functional description to the extent that one can show how the item is organized into the mechanism. Functional description, in short, is a means of integrating an item into a hierarchical nexus of mechanisms. This account (like Craver 2001) is inspired by Cummins' (1975, 1983, 2000) and Toulmin's (1975) discussions of functions. My goal is to situate this "analytic account" with respect to the contemporary mechanical philosophy.

2 What Makes a Neurotransmitter a Neurotransmitter?

Let us begin with a simple and uncontroversial example of a functional description in the neurosciences: the neurotransmitter. To call something a neurotransmitter is to suggest that it is the kind of thing that can be used to send signals from one cell to another. Consider the evidence required to establish that a given chemical substance is a neurotransmitter.

Table 1 lists the six criteria that appear in most introductory neuroscience texts. Although one or more of these criteria are violated for some known neurotransmitters (especially amino acid transmitters like glutamate), they nonetheless represent well the kinds of evidence used to evaluate functional descriptions. Specifically, they are designed to show that the putative neurotransmitter is situated in the mechanisms of the synapse (see Fig. 1) in such a way that it can plausibly act as a means of intracellular communication. The first criterion is that the transmitter must be present in the axon terminal. In Fig. 1, the transmitter is shown stored within circular vesicles floating in the cytoplasm of the axon terminal (C). This criterion is relevant because the axon terminal is the paradigmatic starting point for the

Table 1 Six traditional criteria for identifying a neurotransmitter

1. The chemical must be present in the presynaptic terminal
2. The chemical must be released by the presynaptic terminal in amounts sufficient to exert its supposed action on the postsynaptic neuron (or organ). Release should be dependent upon inward calcium current and the degree of depolarization of the axon terminal during the action potential
3. Exogenous application of the chemical substance in concentrations reasonably close to those found endogenously must mimic exactly the effect of endogenously released neurotransmitter
4. The chemical must be synthesized in the presynaptic cell
5. There must be some means of removing the chemical from the site of action (the synaptic cleft)
6. The effects of the putative neurotransmitter should be mimicked by known pharmacological agonists and should be blocked by known antagonists for that neurotransmitter

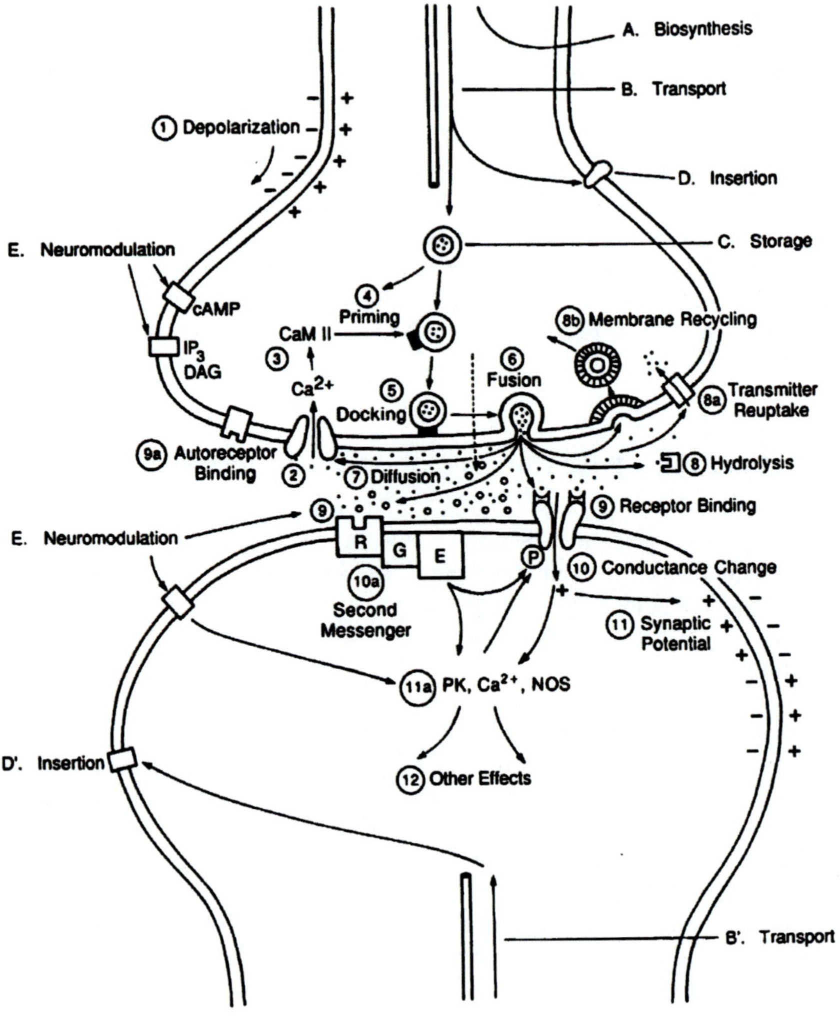

Fig. 1 Mechanisms in the synapse (Reprinted from Shepherd 1994)

mechanisms of intraneuronal communication. The second criterion requires that the substance should be released in a calcium- and depolarization-dependent manner and should be released in amounts sufficient to exert its supposed action on the postsynaptic cell. The release of the substance should be calcium and depolarization dependent because the mechanisms of neurotransmitter release are typically taken to begin with the depolarization of the axon terminal (thus linking the chemical signal across synapses to the intraneuronal electrical signal that the synapse is to preserve). Depolarization typically effects neurotransmitter release by opening voltage-sensitive Ca^{2+} channels in the axon terminal (see Fig. 1; items 1–6). Furthermore, if the substance is to act as a synaptic signal, it must be released under physiologically relevant conditions in sufficient quantities that it can produce effects in the postsynaptic cell (see Fig. 1; items 10–12). The third criterion requires that these postsynaptic effects be produced by both exogenous and endogenous applications of the chemical. This criterion tests the causal relevance of the substance to the activity in the postsynaptic cell. Evidence concerning presynaptic synthesis, the fourth criterion, is required to show that the chemical's production is subject to mechanisms of regulation typical of neurotransmitter systems. The fifth criterion that there is some mechanism for removing neurotransmitter from the cleft is required because the tight relationship between action potentials in the presynaptic cell and transmitter concentration in the cleft would not hold if there were no mechanisms for disposing of "excess" transmitter in a timely fashion (see Fig. 1; items 7, 8, and 8a). Finally, the putative postsynaptic effect of the chemical substance should be mimicked or blocked by pharmacological agents known to activate or impede the postsynaptic receptors for that substance. Again, this criterion is required to test the causal relationship between the presence of the substance in the cleft and the postsynaptic response. If agonists cannot mimic the substance, then one has some reason to doubt whether the substance itself is responsible for the postsynaptic effect. If interfering with postsynaptic receptors does not block the effect, then the substance at least does not act in a manner typical of neurotransmitters.

So what makes a neurotransmitter a neurotransmitter? To presage the discussion of Sect. 6, note that the criteria express no commitments about the developmental or evolutionary origins of the molecule in question. For those who embrace an adaptational view of functions, to claim that a substance has the function of mediating communication between cells (as evidenced by the six criteria) involves asserting (i) that the chemical substance came to be at this synapse because it can mediate communication between cells and (ii) that the chemical substance is capable of mediating communication between cells. Although criteria (1–6) are clearly designed to satisfy some requirement like (ii), precisely none of them address (i). Indeed, it appears that the evidential order of things is the other way around: the evidence for (ii) is the best reason for believing (i). Regardless of how the molecule came to be used as a neurotransmitter (by drift, exaptation, evolution, chance, or divine fiat), so long as it satisfies criteria (1–6), it still functions as a neurotransmitter at the synapse.

Second, to presage the discussion of Sect. 7, criteria (1–6) go well beyond merely exhibiting an input-output relationship. After all, nothing is "put into" or "put out of" the neurotransmitter (except in synthesis and enzymatic degradation). And although some of the criteria (especially 3–6) do address relationships that might be represented in an input-output function, the others do not. Rather, the criteria are designed to show that a given chemical substance is situated among the mechanisms of chemical transmission in such a way that it can fulfill the role of a neurotransmitter. This involves not merely specifying some IO relation of the chemical substance but, in addition, showing that the exercise of the capacities thus described is organized into the mechanisms for regulating the chemical's synthesis, release, and removal from the synapse. Were one merely to describe how neurotransmitters are synthesized or how they bind to postsynaptic receptors, one would not have evidence that the substance functions as a neurotransmitter. The function "neurotransmitter" reaches out into the mechanisms of the synapse to include other details about the mechanisms in the pre- and postsynaptic cell. To describe a neurotransmitter so locally would be like describing a spark plug as having the function of making sparks; it would describe it in isolation from its mechanistic environment.

In short, criteria (1–6) concern neither the history of the substance nor its local interactions with other parts of the cell, but rather how the substance is situated within the mechanism of synaptic communication. To describe a substance as a neurotransmitter is to describe how it fits into a containing system (Cummins 1975) or a mechanism (Toulmin 1975).

3 Mechanisms

But what is a mechanism? This question has received intense philosophical discussion over the last decade (see Machamer et al. 2000; compare, e.g., Bechtel and Abrahamsen 2005; Bechtel and Richardson 1993; Burian 1996; Darden 2006; Glennan 1996, 2002; Salmon 1984; Thagard 2000; Wimsatt 1976). I prefer my own account (Craver 2007), which is a descendant of the account in Machamer et al. (2000) but supplemented with a view of causal relevance owing to Woodward (2003).

Roughly, a mechanism is a collection of entities and activities organized such that they give rise to the behavior of a mechanism as a whole. Entities are objects, such as neurotransmitters and cells. These entities are characterized in terms of structural properties, such as their size, conformation, and material constituents, and in terms of their relations with other entities in the mechanism (their locations, relative motions, forces). Activities are the things the entities do, such as binding to receptors and generating action potentials. Activities, on my Woodwardian interpretation, are typically characterized by a set of generalizations concerning the properties and organizational features required for an activity to occur and the consequences of such occurrence. Such generalizations describe, for example, the properties that are required for different activities to occur (e.g., molecular conformations), the sphere of influence of the activity (e.g., obeying the inverse square law), and its direction of action (e.g., linearly or at right angles). The activities in neuroscience and physiology tend to be mechanistically explicable. The neuronal activities of generating and propagating action potentials, for example, can be explained in terms of the activities of ions and proteins in the cell membrane.

These entities and activities are organized together spatially, temporally, and actively such that they give rise to the phenomenon to be explained. Forms of spatial organization include the size, shape, location, orientation, and compartmentalization of the various parts of the mechanism. Forms of temporal organization include the orders, rates, and durations of the various activities. Active organization is a matter of which entities act and interact with one another, for example, whether they are organized in series or in parallel and whether they involve cycles and feedback loops. Mechanism schemas, texts or diagrams that describe mechanisms at various grains, describe the relevant properties of the entities and detail the overall organization of the mechanism by virtue of which it gives rise to the behavior of the mechanism as a whole. The components are bound together into a single mechanism in part because of the causal interactions among them and, more fundamentally, because of their relevance to the behavior of the mechanism as a whole.

This notion of mechanism, exemplified time and again in contemporary biology, physiology, and neuroscience texts, clearly breaks with the historic association of mechanism with a set of basic and catholic causal principles. The kinds of activities that appear in contemporary mechanistic explanations are far more diverse than austere mechanists would allow. Furthermore, mechanisms need not be deterministic (probabilistic mechanisms are common) or sequential (they may involve feedback, forks, joins, and causal loops). Though descriptions of mechanisms must start and end somewhere, mechanisms themselves might have no clear beginning or end and often run in cycles that are only artificially described as working from start to finish (such as the Krebs cycle or the mechanisms underlying circadian rhythms). This liberalization of the concept of mechanism has expanded the explanatory potential of the mechanical worldview while trading away only the Enthusiasm of austere mechanical philosophers.

As an example of a mechanism, consider how the NMDA receptor/ionophore complex works. This mechanism is named for what it does. It is a receptor because it binds the neurotransmitter glutamate (and pharmacological agents that mimic glutamate, such as *N*-methyl-D-aspartate (NMDA)). It is an ionophore because when it binds to glutamate, it forms an ion channel traversing the membrane of the neuron. Activation of the NMDA receptor is a means of transforming an extracellular chemical signal (born by neurotransmitters) and an intracellular electrical signal (born by ion fluxes in the cell) into an intracellular chemical signal (born by intracellular ions and molecules). It can thus be described as working from beginning to end. The extracellular chemical signal comes in the form of neurotransmitters (glutamate and glycine) that bind to extracellular binding sites. When they so bind, the protein changes its conformation, exposing a channel through its center. Under resting electrical conditions of the postsynaptic cell, the ion channel is blocked by positively charged magnesium (Mg^{2+}) ions held in place by electrical attraction and repulsion. When the postsynaptic cell depolarizes (as when it is in an excited state), the cell becomes less negative (and eventually positive) with respect to the extracellular fluid. The electrical forces holding the Mg^{2+} in the channel weaken, and the Mg^{2+} ions drift out of the channel, allowing Ca^{2+} (the intracellular chemical "signal") to diffuse into the cell.

This brief description includes the entities (e.g., glutamate, binding sites, channels, ions, membranes) and activities (e.g., binding, blocking, repelling, depolarizing) that constitute how the mechanism works. The activities can be characterized in terms of more or less invariant change-relating generalizations specifying, for example, that glutamate binding changes the pore's conformation, that depolarization removes the Mg^{2+} blockade, or that opening the channel allows Ca^{2+} to flow into the cell. The components are organized spatially (e.g., the channel spans the membrane), temporally (e.g., depolarization precedes the release of the Mg^{2+} ions), and actively (e.g., the transmitter binds to the receptor). The organization of these parts gives rise to the behavior of the mechanism as a whole: a highly regulated gating of Ca^{2+} currents across the membrane. One could make the mechanism behave differently or not at all by intervening to change these components or to alter their characteristic organization.

This mechanism (as with all mechanisms as the contemporary mechanical philosophy describes them) is explicitly defined in terms of what it does. The mechanism works from beginning to end, where the end is not what the mechanism invariably does but what we think it is supposed to do. Mutant NMDA receptors, for example, might not work this way, and even perfectly healthy and "normal" NMDA receptors might fail to open under the appropriate conditions if only because the molecular movements involved in channel opening are stochastic. Perhaps it is true that most or all NMDA receptors work in this way, but if so, this is an accidental fact added to the functional description, not something constitutive of its functioning as such. One might describe the behavior of a most irregular mechanism (such as the mechanism of neurotransmitter release) or even a mechanism that has exactly one instance, in exactly the same way. The sense of "normal" here is thus not synonymous with "universal" or "regular" or "typical" but instead should be understood as specifying how the receptor works when glutamate synapses work as *they* normally do and so on, until the hierarchy ends in some behavior that the scientist is interested, for whatever reason, in explaining.

This teleological feature of mechanistic description is also implicit in the fact that mechanisms such as the NMDA receptor mechanism are bounded: a judgment has been made about which entities, activities, and organizational features are in the mechanism and which are not. The world does not come prechunked into mechanisms; it takes considerable effort to carve mechanisms out of the busy and buzzing confusion that constitutes the causal structure of the world. Some mechanisms are entirely contained within physical compartments, such as a nucleus, a cell membrane, or the skin. Transcription (typically) happens within the nucleus, and translation occurs in the cytoplasm. However, mechanisms more frequently transgress compartmental boundaries. The description of the mechanism of the NMDA receptor, for example, relies crucially on the fact that some components of the mechanism are inside the membrane and some are outside. Even the simple act of carving such a mechanism into working parts, as opposed to mere spatiotemporal pieces that might be produced by slicing, dicing, or cubing, requires some principle by which one can recognize a difference between ways of chunking that are relevant to some end and those that are not (Kauffman 1971).

Austere mechanists, eschewing final and formal causes from their explanations in favor of corpuscles operating blindly by motion and contact, lacked principles to define the unity of a machine, organ, or organism. Descartes at times favors principles of spatial organization: the parts are within spatial boundaries, they move together, and they can be transported together from one place to another while maintaining fixed relative positions with the other components (see Des Chene 2001). Others (such as Salamone De Caus) appeal to contact among the parts. Contemporary physiologists recognize counterexamples to each of these suggestions. I have already noted that mechanisms frequently defy tidy physical boundaries (although every mechanism can, trivially, be circumscribed). Parts of mechanisms also often move in separate directions (as any multiple-pulley system illustrates). Some mechanisms are more ephemeral than others; they work only as components happen to come into the appropriate spatial arrangement.[2] For example, in many biochemical cascades, the relevant reactions could happen anywhere in the cytoplasm. Such mechanisms lack stable spatial relations; they could not be picked up and carried from one place to the next. Those accustomed to drawing and studying tidy diagrams of mechanisms in physiology textbooks can temporarily forget that bodies and cells, for example, are bubbling stews of entities and activities and that it takes considerable scientific effort, abstraction, and idealization to distinguish components from contraband, activities from incidental interactions, and causes from background conditions. And this filtering process requires (essentially) fixing on some behavior, process, or *function* for which a mechanistic explanation will be sought (see Craver 2007, Chap. 4).

In a slogan, mechanisms are the mechanisms of the things that they do. The entities and activities that are part of the mechanism are those that are relevant to that function or to the end state, the final product that the mechanism, by its very nature, ultimately produces. Relevance here should be understood in part in terms of relations of mutual manipulability between a component and some behavior of a mechanism that one seeks to understand; in short, a component is relevant to a behavior of the mechanism as a whole if one can manipulate the behavior of the mechanism as a whole by intervening on the component (as in lesion experiments or electrical stimulation) and one can manipulate the behavior of the component by intervening to stimulate or inhibit the behavior of the mechanism as a whole (see Craver 2007, Chap.4, Section 8). Furthermore, we divide a system into parts in part by deciding first what needs to be done in order for the mechanism as a whole to behave as it normally does. The NMDA receptor's function (to turn a joint chemical and electrical stimulus into a change in intracellular Ca^{2+} concentration) determines which features of the channel structure are especially important or necessary for just that role. From the perspectivalist view adopted here, these judgments of normality continue upward until they are grounded ultimately in the judgment or interests of an observer.

[2] See Glennan (2009).

Functions, on this view, are roles within mechanisms, defined ultimately in terms of a topping-off point selected for its relevance to observer interests and perspectives. Mechanisms come into view as entities and activities organized to perform such functions.[3]

4 Levels of Mechanisms

One reason neuroscientists are interested in the NMDA receptor is because its behavior is a component in the mechanism of long-term potentiation (LTP). LTP is one of the means by which certain neurons in the central nervous system (CNS) strengthen their connections (synapses) with one another. When the presynaptic neuron (the one that releases neurotransmitters) and the postsynaptic neuron (the one containing the NMDA receptor) are simultaneously active, the synapse is strengthened (LTP is "induced"). When the presynaptic neuron is active, it releases glutamate (and glycine) into the synaptic cleft. The postsynaptic cell is active when it is depolarized from its resting electrical potential. These two factors, recall, are the crucial set-up conditions for the opening of the NMDA receptor. The termination condition (the influx of Ca^{2+}) is a crucial stage in the induction of LTP. Many neuroscientists believe that LTP is a crucial activity in the mechanisms of some kinds of learning and memory. For example, LTP is a component in the mechanisms of spatial map formation in the hippocampus (a medial temporal lobe structure), and these spatial maps are thought to be components in the mechanisms of spatial memory, the ability to learn to navigate through novel environments. The NMDA receptor (an entity) and LTP (an activity) are also thought to be involved mechanisms that "top off" in drug addiction (Kauer and Malenka 2007) and Alzheimer's disease (Rowan et al. 2003), which could in no compelling sense be described as adaptations.

All these theories describe mechanisms at multiple levels of organization. The activity of the NMDA receptor is of interest by virtue of the fact that it is a stage in the mechanisms of LTP induction, which constitutes a stage in the mechanism of spatial map formation, and so on. Such multiply embedded hierarchies are usefully thought of as levels of mechanisms: they are part-whole relations with the additional restriction that the parts are components organized together to produce the behavior of the mechanism as a whole. To be at a lower (−1 or $-m$) level just is to be one of the components organized into the mechanism as a whole, which constitutes the higher (+1 or $+n$) level. Of course, there are other useful notions of "level" in biology (tracking, e.g., objects of different sizes, the phenomena in different theories, the domains of different sciences, and the targets of different techniques). However, levels of mechanisms capture a common notion of level, and one that is especially relevant to thinking about the relationship between functions and mechanisms in physiological sciences.

[3] Given the hierarchical embedding of mechanisms to be discussed below, functional description is often appropriate both for the behaviors of mechanisms as a whole (either because they have been privileged as such by an observer or because they play a role in a higher-level system that is so privileged) and for the roles of the parts in producing that behavior.

Different scientists top off their mechanistic hierarchies in different highest-level activities. Some biologists are interested in, for example, channel structure and how channels open and close. They are not especially interested in system-level mechanisms or cognition. Some are interested in cognition, others in social bonding, and others in ecological systems. Sometimes biologists direct their attention toward mechanisms and functions that contribute to an organism's fitness. Sometimes they want to know how diseases work, how toxins kill cells, or how pollutants change the dynamics of an ecosystem. Differences in topping-off points reflect differences in interest and emphasis, and these differences are reflected in the mechanistic theories that different scientists, fields, or traditions use to explain the phenomena in their domain. It is by reference to these historically, individually, and disciplinarily relative topping-off points that the relevance of lower-level components is determined. The choice of a topping-off level selectively focuses the researcher's attention upon certain lower-level mechanisms and not others. It can lead researchers to carve the system into altogether different parts, as Kauffman (1971) and Wimsatt (1974) emphasize. In our working example, an antecedent interest in spatial memory focuses the investigator's attention upon the mechanisms of spatial map formation, LTP, and the mechanisms of NMDA receptor activation. The mechanisms of NMDA receptor activation have also been hypothesized to play crucial roles in hierarchies that top off in the progression of Huntington's, the psychological effects of PCP abuse, the mechanisms of programmed cell death, and the mechanisms of chronic pain. The choice of a topping-off point is a crucial step in filtering the causal nexus to yield a properly mechanistic nexus. This is why I am a perspectivalist about functions and mechanisms.

I hasten to emphasize that this perspectivalism has limits. Ultimately, the causal structure of the world, facts about what variables make a difference to which others and which entities and activities exist and occur, allows only some perspectives to fit. It is an empirical question whether a system exhibits the behavior that one is trying to explain. It is an empirical matter whether a given entity, activity, or organizational feature exists and whether it is in fact relevant (in the sense sketched above) to the phenomenon thus described. My point is that the actual causal structures of the body, the brain, and the cell are bewilderingly complex and reticulate. This is why it is such a significant scientific achievement (Haugeland 1998) to properly characterize a function and to generate a multilevel description of mechanisms that accommodates all of the data about the parts, activities, and organizational features at multiple levels and weaves them into a coherent image of how something works. That said, there are many ways of decomposing such bewilderingly complex bits of the causal nexus into intelligible units, and the identification of functions and mechanisms is crucial for bringing intelligible order to such a causal stew. They are crucial steps, that is, in providing explanations.

5 Explanation: The Mechanist's Stance

Logical empiricist philosophers of science (such as Hempel 1965) once thought that explanations are arguments showing that a description of the phenomenon to be explained follows from statements describing covering laws and relevant conditions.

This elegant and powerful view faded from currency because arguments and explanations have different criteria of adequacy; inferential subsumption under general laws is neither necessary nor sufficient for an adequate explanation (see Salmon 1989; Craver 2007, Chap. 2). Rather, to explain a phenomenon is to show how it is situated in the causal nexus (Salmon 1984). More plainly, explanations reveal the causal structure of the world. I embrace this view of explanation, but not Salmon's view of the causation.

For Salmon, the causal nexus is composed of causal processes (understood as space-time continuants bearing conserved quantities) that interact with one another when they intersect one another in space-time and exchange conserved quantities (Salmon 1994; Dowe 2000). This view embodies the boldness and simplicity of the earliest statements of the mechanical philosophy, but it is not ideal for thinking about sciences such as neuroscience. The process view emphasizes relatively fundamental kinds of causal interaction (e.g., those that involve collisions or charges). Physiological activities such as the opening of an ion channel, the transcription of DNA, and the formation of spatial maps are much too complex for tidy description from this perspective. Furthermore, the process view requires that causal processes intersect one another in causal interactions. This means that the causal nexus so conceived has no room for causation by omission or double prevention (e.g., inhibiting an inhibitor), forms of causation that are literally ubiquitous in the physiological sciences (for a fuller discussion, see Craver 2007). Depolarization of the postsynaptic cell causes Ca^{2+} to enter through the NMDA receptor, but it does so by removing a process (Mg^{2+} ions) that prevents the ions' flow.

So we must understand the causal nexus more liberally in physiological sciences. As discussed briefly above, the interactions and activities (following Woodward's (2003) account of actual causation) can be understood in terms of relations by which the value of one variable (standing for a property, or the presence or absence of an object, or the occurrence or nonoccurrence of an activity) depends upon the value of another. Such dependency amounts to the fact that one can change the value of the first variable by intervening to change the value of the second (given certain restrictions on the intervention). On this view, the causal nexus can be represented roughly as a set of variables related by generalizations that remain stable (or invariant) when one intervenes to change the value of the variables in the generalization. Such a view explicitly allows for causal interactions above the fundamental level (as there is no metaphysical restriction on the kinds of objects or properties that might enter into a causal relationship) and has no difficulty accommodating causation by omissions and preventions. This view also comports nicely with the kinds of experiment one uses to test causal claims. One intervenes to change the putative cause and detects the changes, if any, in the effect variable under those controlled conditions. Explanation on this view is a matter of revealing causal dependency relations of this sort or, for the explanation of singular events, tracing the productive relations among the entities and activities that make such change-relating generalizations true. Either way, this view remains true to Salmon's

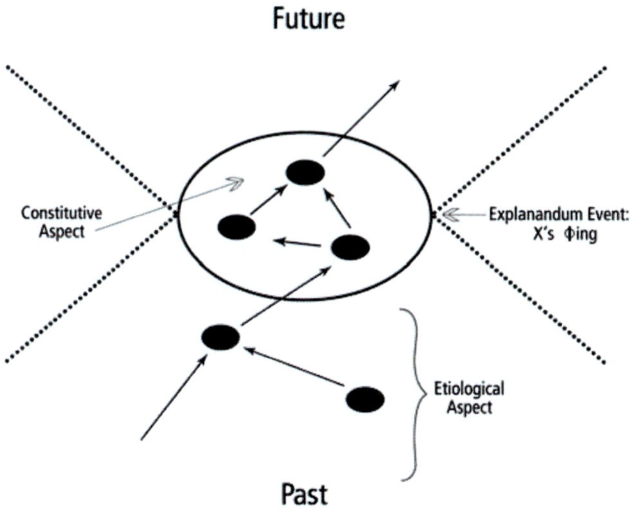

Fig. 2 Constitutive and etiological aspects of causal-mechanical explanation

overall vision: to explain an event or phenomenon is to show how it is situated in the causal structure of the world.

Salmon recognizes two ways of situating a phenomenon in the causal structure of the world: an etiological form of explanation, in which one explains a phenomenon by tracing its antecedent causes, and a constitutive explanation, in which one explains a phenomenon by revealing its internal causal structure. In order to accommodate the diverse explanatory roles played by functional description in physiological sciences, it is necessary to add a third contextual variety of causal-mechanical explanation, further liberating the contemporary mechanical philosophy from its historical strictures and recognizing within that philosophy an essential place for functions in our effort to make the causal structure of the world intelligible.

6 Etiological Explanation and Adaptational Functions

Etiological explanations are typically offered in response to questions concerning the origins of some item, its path of development, or its historical trajectory. Salmon represents this etiological aspect of mechanistic explanation in the bottom portion of Fig. 2. The figure illustrates the *backward-looking* character of etiological explanations; such explanations highlight the pathway connecting relevant set-up conditions in the past, through intermediate stages of activity, to the item to be explained.

Some philosophers and scientists reserve the term function for traits, properties, and activities that are adaptations (Ruse 1971; Wimsatt 1972; Brandon 1990). Churchland

and Sejnowski claim that this use captures the sense of "function" used in neuroscience (1992, 69; see also Bechtel 1989).[4] I focus on Wright's classic formulation:

The function of *X* is *Z* means:

(a) *X* is there because it does *Z*.
(b) *Z* is a consequence (or result) of *X*'s being there (1973, 161).

In the standard biological case, (a) is embellished as a natural-selection story roughly to the effect that heritable traits of type *X*, by virtue of their doing *Z*, increased the likelihood that organisms bearing traits of type *X* would survive and/or reproduce and as a result contributed to the preservation of traits of type *X* in a given population. The NMDA receptor has the adaptational function of mediating cellular "signals" if and only if the NMDA receptor allows these signals to be mediated and was preserved in organisms because it did so in the past. One advantage of identifying biological functions with adaptations is that doing so can often accommodate the intuition that a trait's function explains its presence. Asked why the mouse has NMDA receptors, it may be correct to respond that the NMDA receptors are there *because* they mediate certain chemical signals. The "because" in this sentence is the "because" of efficient causation: adaptational explanation is an example of the *etiological* type of mechanistic explanation.

The ability to make sense of this kind of functional explanation is a notable advantage of causal-mechanical models of explanation relative to covering-law models. The puzzle of functional explanation is to accommodate the intuition that a trait's presence can be explained by appeal to what it does. We explain why we have NMDA receptors by appeal to their role in learning and memory, for example. A defender of the covering-law model of explanation faces the challenge of showing that one can derive or otherwise infer the presence of a particular type of trait from what the trait allows the organism to do. However, as Hempel recognized, the fact that the same function can be produced by multiple functionally equivalent types of trait always brings the explanatory argument up short: at most one can infer that one of the functional traits exists, not the functional type that one seeks to explain. The causal-mechanical view, on this understanding, thus proposes to show that the surface character of functional explanation (that an item's presence can be explained by adverting to its function) can be translated into an etiological framework that describes a selective developmental or evolutionary process. One describes the mechanisms beginning with the first appearance of the trait and ending with its contribution to survival and reproduction. This is well and good, so long as we bear in mind that the translation is only approximate. The presence of an item in an organism now is explained in terms of the behavior of items of the same type in the past. The token effects of the trait now, however, do not explain the presence of the item. The teleology preserved in the assimilation of functions to adaptations thus breaks with the Aristotelian idea of a goal or purpose as a cause of behavior over and above its constitutive and etiological explanations. (There is reason to doubt

[4] They also note that the function of an item is its "job" and that any apparent teleology in the sense of function is "eliminable or reducible without remainder in an evolutionary framework." For classic adaptational accounts, see Brandon (1990), Millikan (1984), Neander (1991), Ruse (1971), Wimsatt (1972), and Wright (1973). Garson (2008) provides a recent review.

that Aristotle held this view of final causes; see Leunissen 2007.) The assimilation of functional language to language about kinds of causal histories on this understanding is eliminative, not reductive. To ascribe a function is a shorthand way of describing it as having a certain kind of history and, one might say, nothing more than that. The adaptational function of an item makes no further contribution to how an item is situated in the causal structure of the world.

Functional descriptions, however, are sometimes said to have a normative dimension (see, e.g., Neander 1991; Wimsatt 1972; Wright 1973). The functional description distinguishes an item's preferential behavior (its *proper function*) from the item's myriad nonfunctional effects in a system (e.g., a receptor deforms the lipid bilayer of the cell membrane), the myriad things that the item might function *as* (e.g., a target for pharmacological intervention), and the many ways that an item might malfunction (e.g., a mutation causes a receptor not to bind with a neurotransmitter). Functional descriptions thus conceived describe how things ought to work rather than how they in fact work. This way of thinking about functions fits naturally with the idea that creatures have been designed; the function of the receptor is the purpose for which the demiurge created it and arranged it just so. And perhaps this manner of speaking can be translated into our posttheistic biology by putting evolution by natural selection in the role of a divine maker: roughly, an item's function is that effect in virtue of which its type has been preserved in a species. Thus, Wimsatt claims, "Given the operation of differential selective processes, it is possible to show that any given system resulting from this process has all the relevant logical features of purposiveness and teleology" (1972, 16). Selective processes, it is said, define goal states within higher-level systems or preferable states of individual or species-level traits (Hull 1974). If so, one can reduce facts about how a physiological item *ought to* behave to facts about how items of that type *do* or *did* behave.

To make good on the proposed reduction, one should be able to form an argument that begins with premises describing the selective history of an item and concludes with statements about its goals, purposes, and preferential states. However, I know of no successful argument that begins with premises about what causes what and ends with conclusions about what ought to be the case: either the "ought" is smuggled into one of the premises or the proposed derivation relies on obvious "tricks" peculiar to formal logic or the notion appears from out of the blue in the conclusion (see Russell 2010 for a critical examination of some clever attempts to derive an ought from an is). Machamer (1977) reconstructs the proposed argument linking facts about an item having been selected for some behavior to conclusions stating that the item's behavior is good, preferable, or a goal state. He demonstrates convincingly that this argument succeeds only if one tacitly presumes the existence of a higher-level containing system, the behavior of which is good, preferable, or a goal state. One must presume, for example, that it is preferable that an organism should live and reproduce, that the species ought to survive, or that one ought to live some conception of the good life. But these shoulds and oughts, on the perspectivalist view recommended here, are ultimately projections of our interests or preferences. The causal structure of the world does not ground talk of goals, purposes, and preferential states. Such things are "queer" in a mechanistic world (in Mackie's phrase; 1977) because they are "fraught with ought" to borrow from Sellars (see 2007).

Schaffner (1993) develops a line of arguments suggesting that selection is neither necessary nor sufficient to make talk of goals, purposes, and preferential states appropriate. One can understand the growth of thunderstorms, the fine-grained sand on a beach, and the momentum of a pachinko ball at the bottom of a pachinko machine in terms of selection mechanisms (see Schaffner's cloner example), but few are willing to assign functions (in a nonperspectival sense) to thunderstorms, the fineness of a grain of sand on the beach, or the momentum of a pachinko ball. Further, as has often been noted, the etiological reduction of functions prevents one from assigning functions to traits on which selection has not yet acted. The first NMDA receptor did not have a function according to this view because a trait can have a function only in the second generation (after it has contributed to fitness) and only if the first NMDA receptor in fact manifested its dispositions in a way that contributed to the organism's reproduction. The perspectivalist avoids this consequence.

None of this is news. I emphasize it not because evolutionary thinking is out of place in physiological sciences such as neuroscience. Indeed, in trying to find an intelligible picture of what organisms are doing and how they do it, it is most useful to consider the selective forces that have likely shaped their development. (I say likely because of phenomena such as drift and exaptation; see Gould and Lewontin 1978; Gould and Vrba 1982.) Evolutionary thinking can be heuristically useful as a guide to creative thinking about what an organism or organ is doing, the conditions under which it is suited to work, and about its apparent failure to work optimally, as one would expect had it been created by a benevolent, omniscient, and omnipotent designer.

I reiterate these objections to the normative implications of selective etiologies for two reasons. First, neuroscience and physiology have goals that would be hampered by the general acceptance of such a proprietary notion of function. Much of physiological science such as neuroscience is driven not by the goal of understanding how the nervous system functions when it is working properly but rather by the goal of understanding how it can fail and how such failures might be predicted and controlled. One can describe the function of items in the mechanisms for anoxic cell death, the production of cancer, and the progression of Alzheimer's disease. One can describe the function of items in the mechanisms for anoxic cell death, the production of cancer, and the progression of Alzheimer's disease. When one describes an oncogene as an oncogene, one is describing it functionally without being committed to the idea that the oncogene survived by virtue of being an oncogene. Indeed, it would seem likely that it survived in spite of the fact that it functions as an oncogene. Likewise, a researcher hoping to build a robotic interface with someone's motor cortex, for example, might hunt for signals that can be commandeered for the purpose of moving the arm even if such signals were not at all part of anything that evolution by natural selection might have considered. These are as much a part of the mechanisms of the brain as are those parts and mechanisms that have been selected for their effects. Researchers approaching the brain from such a translational perspective will see functions where the advocate of adaptational functions does not.[5]

[5] Analyses of the concept of function in terms of current ability or propensity to survive and reproduce (e.g., Bigelow and Pargetter 1987; Boorse 1976; Canfield 1964) likewise fail to accommodate many of the perfectly legitimate uses of functional language that can be found in neuroscience.

Second, if anyone has an obligation to think slowly and pedantically about the normative implications of evolutionary biology, it is philosophers of biology. Those who claim to find in evolutionary biology a means of distinguishing the good from the bad, the healthy from the diseased, and the deleterious from the beneficial owe a compelling argument for the ability of evolutionary processes to ground such conclusions. The perspectivalist refuses to pass the buck for these normative judgments to evolutionary biology. When we make such normative judgments, says the perspectivalist, we are responsible for justifying them. Evolution cannot do that work for us.

7 Constitutive Explanations and IO Functions

Consider a second role played by functional description in mechanistic neuroscience and physiology. According to this view, a function is a mapping from inputs to outputs in conformity with a rule. Call these input-output (IO) functions. Sometimes Cummins describes functions this way. For example, he says that functions are capacities, where capacities are, "specified by giving a special law linking precipitating conditions to manifestations— i.e., by specifying input-output conditions" (1983, 53). IO functions characterize the activity of some item without reference either to its context or to its internal complexities. In forming such a description, one draws a conceptual dividing line at the spatial boundary of the object or activity and recognizes a limited number of specific *interfaces* across that boundary—more or less well-defined interactions with items outside of that boundary (see Haugeland 1998). For example, in describing the IO function of the NMDA receptor, one begins by parsing it from its environment at its spatial boundary and characterizing the relevant interfaces (interactions) across that boundary. Three significant interfaces between the NMDA receptor and its environment are the binding of glutamate and glycine to the receptor, the blocking action of Mg^{2+} ions, and the influx of Ca^{2+} through the channel pore.

IO functions can sometimes be characterized mathematically. Two examples are represented graphically in Fig. 3a, b (taken from Mayer et al. 1991). The first figure depicts a typical dose-response curve relating concentrations of agonists (glutamate and pharmacological agonists) to the current of Ca^{2+} flowing into the postsynaptic cell in the absence of Mg^{2+}. The second of these characterizes changes in that influx of Ca^{2+} as a function of postsynaptic depolarization in a medium with extremely high concentrations of Mg^{2+}. Both can be understood, as suggested above, as causal generalizations that are invariant under interventions: one intervenes to change the concentration of neurotransmitter in the synapse and detects changes in the current flowing through the channel, as in Fig. 3a, or one holds Mg^{2+} concentrations constant while varying the membrane voltage and recording the current through the channel, as in Fig. 3b. Clearly neither of these invariant change-relating generalizations (cf. Glennan 2002; Craver 2007) characterize completely the activation of the NMDA receptor (i.e., its function), and each characterizes it only under highly constrained conditions (e.g., experimentally gerrymandered levels of Mg^{2+}). Rather, these IO functions and others like them combine to form a complex description of the behavior of the NMDA receptor.

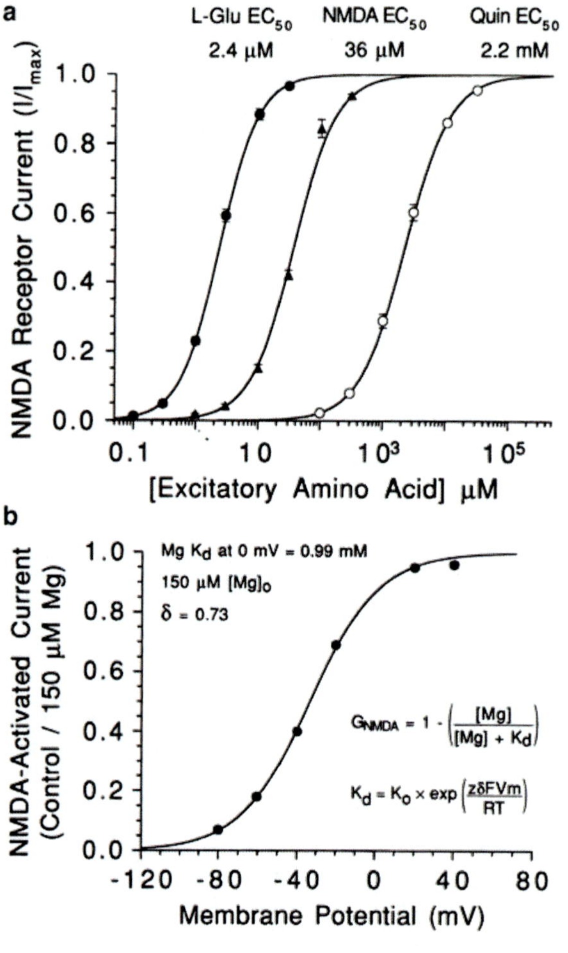

Fig. 3 (**a**) NMDA receptor current as a function of concentration of excitatory amino acids (Reprinted from Mayer et al. 1991) and (**b**) NMDA activated current as a function of membrane potential (Reprinted from Mayer et al. 1991)

This *complex IO function* plays two crucial roles for the physiological scientist beyond providing a precise characterization of the phenomenon. First, such abstract description affords the scientist descriptive leverage over the messy details of the constitutive mechanism that produces the complex IO function. One can speak of the activation of the NMDA receptor without going into the complex and poorly understood details of protein chemistry, and one can speak of LTP induction without detailing the intricate pattern of molecular activities responsible that induce LTP. IO functions are also descriptive tools for dealing with the multiple realizability of most biological functions: that is, for dealing with individual, strain, and species differences. The same IO function might be instantiated by a number of different mechanisms.

Complex IO functions are also important for characterizing the phenomena for which one will seek *constitutive explanations*, the second aspect of causal-mechanical explanation that Salmon recognizes (cf. Bechtel and Richardson 1993). In constitutive (as opposed to etiological) explanations, one explains an event by revealing its internal causal structure. Instead of revealing the causes by which the NMDA receptor is activated, developed, or evolved (as in etiological explanations), one describes the relevant causal structure internal to NMDA receptor, the entities, activities, and organizational features by virtue of which it activates when neurotransmitters are present and the postsynaptic cell is depolarized. Such explanations have been called explanation by decomposition, functional analysis, and explanation by reverse engineering. Constitutive explanations are downward looking in the sense that they describe the internal mechanisms—organized lower (–1 or –*m*) level activities and entities—by virtue of which some aspect of the complex IO function is produced. They situate an item in the causal nexus by detailing the lower-level mechanism that produces those aspects of the complex IO function. The ellipse in the center of Fig. 2 represents this type of explanation. Constitutive explanations are sought when one wants to know how something works or wants to know the "hidden" mechanism by virtue of which an item does something of interest. The explanation of the opening of the NMDA receptor in Sect. 3 is an example of this constitutive form of mechanistic explanation. That explanation is tailored to account for the IO functions represented in Fig. 3a, b and the myriad others like them. It is in this sense that the complex IO function frames the constitutive explanation; they define the relevant input-output relationships that the internal mechanism must be capable of performing.

The language of inputs and outputs that characterize the behavior of the NMDA receptor does not apply straightforwardly to the example of neurotransmitters with which we began. Neurotransmitters are not mechanisms for transforming inputs into outputs, at least as commonly conceived. One can describe the synthesis and release of neurotransmitters this way, and one can describe their effects on postsynaptic receptors this way (as in Fig. 3a), but the molecule itself seems to be a passive participant in these change-relating generalizations that describe how the molecule is situated as a component within a higher-level mechanism. Perhaps one could characterize features of the molecule's environment, such as temperature or pH as inputs, and one could characterize the molecule's conformation as an output. But this manner of speaking is strained and to my knowledge would not be adopted by scientists. The function of the neurotransmitter, in other words, is primarily understood contextually.

8 Contextual Functions

Consider four ways of describing the heart's role in the circulatory system. The heart:

(i) Distributes oxygen and calories to the body
(ii) Pumps blood through the circulatory system
(iii) Expels blood
(iv) Contracts

Descriptions (i)–(iii) are contextual (or "wide") in varying degrees; they each describe things that the heart could not do by itself without being organized together with other entities and/or activities. The heart cannot expel blood (iii) without blood, and the expulsion of blood will only circulate it (ii) if the veins and arteries are appropriately organized. Even then, the heart cannot distribute oxygen and calories (i) in the absence of oxygen and calories. A description of the heart's mechanistic role function is contextual to the extent that it makes explicit reference to objects other than the heart itself and its parts. Reference to objects beyond the boundaries of the heart, notice, is not required in describing (iv) the heart's contraction. In describing the heart as contracting, one makes no implicit commitments concerning the mechanistic context in which this activity is embedded. One offers an isolated description of the sort described in the preceding section.

The same can be said of our description of neurotransmitters. Glutamate, for example, might be described as a molecule that:

(i*) Mediates spatial cognition
(ii*) Carries a chemical signal
(iii*) Binds to a postsynaptic receptor
(iv*) Has a characteristic primary sequence and conformation

Again, descriptions (i*) to (iii*) are wide. When one speaks of dopamine or serotonin as neurotransmitters that regulate emotion, control movement, or underlie addiction, one describes the molecule contextually as a component in a larger system. Contextual descriptions of this sort describe some part and its activities in terms of the contribution it makes to a higher (+1 or +*n*) level mechanism. Such descriptions tacitly refer to the fact that if one were to, for example, intervene to change neurotransmitter levels, one could influence the behavior of such higher-level systems. Cummins writes that, "to ascribe a function to something is to ascribe a capacity to it that is singled out by its role in an analysis of some capacity of a containing system" (Cummins 1983, 99), and we should add that functional characterizations often describe those capacities in a manner that includes wider and wider regions of the causal structure of the system under consideration, as in items (i)–(iiii). There is a difference, after all, between knowing that spark plugs produce sparks and knowing how that sparking is situated in the mechanisms of an engine. In the former case, we describe the spark plug's IO function; in the latter we describe its role contextually. Contextual functions are not simply capacities (IO functions) *picked out* by their place in a higher-level mechanism; rather, they are descriptions of the activity of some item in terms of how it is organized into the workings of a higher-level mechanism. One and the same token sparking of a spark plug may be said to be an instance of sparking, of igniting an explosion, of pushing a piston, and of turning the drive shaft depending on how much of the item's context in the causal nexus one includes in the description. There is no firm dividing line between IO functions and role functions; the distinction depends upon where one draws the boundary lines around an object. My point here is that contextual descriptions are invariably richer than their IO counterparts, making clear how a given IO function is situated in some other system that we care about.

Contextual, isolated, and constitutive descriptions should not be seen as corresponding to divisions in the furniture of the world. They should rather be thought of as distinct perspectives on a hierarchy of levels of mechanisms. As Lycan puts it, "See Nature as hierarchically organized in this way and the 'function/structure' distinction goes relative: something is a role as opposed to an occupant, a functional state as opposed to a realizer, or vice versa, only modulo a designated level of nature" (1987, 78; cf. Churchland and Sejnowski 1992, 18–27). I put it like this: see the world as a mechanistic hierarchy and the distinction between a contextual (+1 or +*n*) function, an isolated behavior (0), and its constitutive (−1 or −*m*) mechanism goes relative to a perspective on level in that hierarchy.[6] One cannot describe an item's role, in the broad sense intended here, without describing the place of its IO function in some more inclusive mechanism.

There is thus a need to recognize a third form of mechanistic explanation beyond those recognized by Salmon: *contextual explanation* (Craver 2001). Sometimes a neuroscientist or physiologist is ignorant of what a given item does or is good for, and this leads her to search for a higher-level mechanism within which it has a role. The answer to such a request for explanation comes in the form of a description of how an item is situated in a higher-level mechanism. The process of situating an item in a higher (+1 or +*n*) level mechanism involves showing how it is organized (spatially, temporally, and actively) into the higher-level mechanism. Contextual explanations are characteristically outward looking and upward looking. They are outward looking because they refer to components outside of the item to be explained, and they are upward looking because they contextualize that item within the behaviors of a higher-level mechanism. Mechanistic explanation, at least as we now understand it, is thus not synonymous with downward-looking, reductive explanation (though constitutive explanations are reductive in the sense that they explain wholes in terms of parts); there are also upward-looking mechanistic explanations.

To return to the example with which began, the neurotransmitter has to be released in correlation with the electrical properties of the cell, has to be cleared from the cleft, has to act on postsynaptic receptors, and has to exhibit the kinds of active organization within a mechanism revealed by the other criteria in Table 1. The concept of a neurotransmitter, as one of our well-articulated concepts in contemporary neuroscience, provides a model of a contentful functional ascription and

[6] Stephen Toulmin makes this point most beautifully: "There is no clear division of natural processes in the real world, into 'functions' on the one hand and 'mechanisms' on the other. Rather, we draw a distinction between the functional and mechanistic *aspects* of any natural process, in one context or another; and whatever can be viewed as a mechanism, form one point of view and in one context, can alternatively be seen as a function, from another point of view or in another context. Indeed, the very *organization* of organisms—the organization that is sometimes described as though it simply involved a 'hierarchy' of progressive larger structures—can be better viewed as involving a 'ladder' of progressively more complex systems. All of these systems, whatever their levels of complexity, need to be analyzed and understood in terms of the functions they serve and also of the mechanisms they call into play. And when we shift the focus of our attention from one level of analysis to another—from one fineness of grain to another—even those very processes which began by presenting themselves to us under the guise of 'mechanisms' will be transformed into 'functions'" (Toulmin 1975, 53).

of the kinds of evidence by which such ascriptions are to be evaluated. It helps to show precisely what is unsatisfying about glib comments that, for example, dopamine is a happiness neurotransmitter. For in the first place, dopamine does many things in the brain, and associating it with just one of those functions already represents a perspectival simplification of how dopamine fits in the causal nexus of the nervous system. In the second place, such a description implicitly appeals to complex higher-level-mechanisms-we-know-not-precisely-what. One might know that one can regulate one's emotional state by regulating dopamine levels and remain largely ignorant of the complex mechanisms by virtue of which that effect is mediated. We add content to such terse and gestural functional descriptions by revealing the entities, activities, and organizational features by which dopamine contributes to the regulation of emotions. This perspective on functional attribution suggests a regulative ideal in formulating functional attributions: they are contentful and precise to the extent that they explicitly make claims about how an item is situated in its causal context. It is by reference to the evidence for such organization (as is the case for neurotransmitters) that functional attributions are evaluated. Similar remarks apply to common ways of talking about brain regions and genes, for example. When one talks about a gene for aggression or a brain region for decision-making, one is speaking gesturally about how an item fits into a higher-level mechanism, and we make progress in fleshing out the content of such gestures to the extent that we build descriptions of how the item is situated in the causal nexus.

9 Conclusion

Three explanatory perspectives are illustrated in Fig. 4, each of which should be acceptable to those who embrace the mechanistic philosophy as we now know it. The figure, which is intended to replace Salmon's useful diagram of aspects of causal-mechanical explanation, depicts two levels (the top and bottom circles) in a mechanistic hierarchy flanking a complex IO function in the middle. The past and future portions of the causal nexus are to the left and right of the hierarchy, respectively. For each type of explanation, the explanandum is some aspect of the complex IO function in the middle; call it *E*.

Etiological explanations trace the pathway of entities and activities terminating in *E*; they explain how *E* came to be there, came to pass, or came to have some property. Such an explanation is shown on the left-hand side of Fig. 4. It is represented as a single level for simplicity, though any complex etiological explanation will typically span multiple levels as well. Explanation in terms of natural selection is a type of etiological explanation, one that requires an understanding of genes, organs, organisms, populations, and ecosystems. Adaptational explanations are *backward looking*. They are also legitimate answers to a causal reading of the question "Why is E there?"

Constitutive explanations explain how *E* works. They are *downward looking* in that they situate *E* with respect to the portion of the causal nexus at a lower (−1 or −*m*) level in a hierarchy of mechanisms. *E* is a "black box," but if we look within, we find

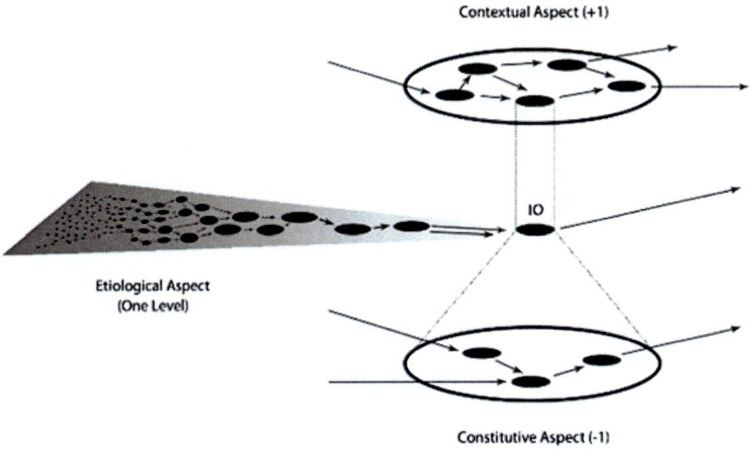

Fig. 4 Constitutive, contextual, and etiological aspects of causal-mechanical explanation

that it is composed of the entities and activities at that level. Complex IO functions are especially useful for describing E without reference to such messy details, but they also frame internal mechanistic explanations; it is a requirement on the adequacy of such explanations that they account (more or less) for the input-output functions of the mechanism as a whole.

Finally, contextual explanations are *upward looking*; they situate *E* with respect to the portion of the causal nexus in a higher (+1 or +*n*) level in a hierarchy of mechanisms. This is why it is explanatory to cite *E*'s role function; contextual role descriptions provide a more or less terse description of how *E* is related to the other entities and activities in a higher-level mechanism. They are therefore legitimate answers to a second reading of the question "Why is E there?" in that they show what the item does as a component in a higher-level mechanism.

In the contemporary mechanical philosophy, functional and mechanistic descriptions work in tandem to bring intelligible order to complex systems. By identifying functions within such systems, one approaches the system with some set of interests and perspectives in mind. One might be interested in understanding how parts of organisms work, how they break or become diseased, or how they might be commandeered for our own purposes. Regardless of which perspective one takes, the identification of functions is a crucial step in the discovery of mechanisms. We no longer speak of mechanisms simpliciter, but rather as mechanisms *for* some behavior. Mechanistic descriptions thus come loaded with teleological content concerning the role, goal, purpose, or preferred behavior of the mechanism. This teleological loading cannot be reduced to features of the causal structure of the word, but it is ineliminable from our physiological, and particularly neural, sciences, precisely because their central goal is to make the busy and buzzing confusion of complex systems intelligible and, in some cases, usable.

Daniel Dennett (1987) suggests that we make the world intelligible by taking different stances: the intentional stance, the design stance, and the physical stance.

My discussion has been about three ways of making things intelligible within a kind of mechanistic design stance, liberated from Dennettian associations with adaptationism and optimality: a stance that there is a behavior that the mechanism as a whole exhibits (that it is the mechanism *of* a behavior) and that the components of the mechanism are organized and interact such that they exhibit its overall behavior. Whether the teleology of our contemporary mechanical worldview is ultimately reducible to features of the causal structure of the world thus depends on whether the ability to *take* a stance with respect to a system can be situated without remainder within the causal structure of the world. And here we have a, perhaps *the*, central puzzle that any properly mechanical understanding of mind must someday face.

References

Allen, G. 2005. Mechanism, vitalism and organicism in late nineteenth and twentieth-century biology: The importance of historical context. *Studies in History and Philosophy of Biological and Biomedical Sciences* 36: 261–283.

Bartel, D.P. 2004. MicroRNAs: Genomics, biogenesis, mechanism, and function. *Cell* 116: 281–297.

Bechtel, W. 1989. Teleological functional analyses and the hierarchical organization of nature. In *Teleology and natural science*, ed. N. Rescher, 26–48. Lanham: University Press of America.

Bechtel, W., and A. Abrahamsen. 2005. Explanation: A mechanistic alternative. *Studies in History and Philosophy of Biological and Biomedical Sciences* 36: 421–441.

Bechtel, W., and R. Richardson. 1993. *Discovering complexity*. Princeton: Princeton University Press.

Bigelow, R., and R. Pargetter. 1987. Functions. *Journal of Philosophy* 84: 181–197.

Boorse, C. 1976. Wright on functions. In E. Sober 1984, 369–385.

Brandon, R. 1990. *Adaptation and environment*. Princeton: Princeton University Press.

Burian, R.M. 1996. Underappreciated pathways toward molecular genetics as illustrated by Jean Brachet's cytochemical embryology. In *The philosophy and history of molecular biology: New perspectives*, ed. S. Sarkar, 67–85. Dordrecht: Kluwer.

Canfield, J. 1964. Teleological explanation in biology. *The British Journal for the Philosophy of Science* 14: 285–295.

Churchland, P.S., and T.J. Sejnowski. 1992. *The computational brain*. Cambridge, MA: MIT Press.

Craver, C.F. 2001. Role functions, mechanism, and hierarchy. *Philosophy of Science* 68: 53–74.

Craver, C.F. 2002. Structures of scientific theories. In *The Blackwell guide to the philosophy of science*, ed. P.K. Machamer and M. Silberstein. Malden: Blackwell.

Craver, C.F. 2007. *Explaining the brain: Mechanisms and the mosaic unity of neuroscience*. Oxford: Clarendon.

Craver, C.F., and L. Darden. 2001. Discovering mechanisms in neurobiology: The case of spatial memory. In *Theory and method in the neurosciences*, ed. P.K. Machamer, R. Grush, and P. McLaughlin, 112–137. Pittsburgh: University of Pittsburgh Press.

Cummins, R. 1975. Functional Analysis. *Journal of Philosophy* 72: 741–764. Repr. In E. Sober 1984, 386–407.

Cummins, R. 1983. *The nature of psychological explanation*. Cambridge, MA: Bradford/MIT Press.

Cummins, R.E. 2000. 'How does it work' versus 'what are the laws?' Two conceptions of psychological explanation. In *Explanation and cognition*, ed. F. Keil and Robert A. Wilson, 117–145. Cambridge, MA: MIT Press.

Darden, L. 2006. *Reasoning in biological discoveries*. New York: Cambridge University Press.

Dennett, D. 1987. *The intentional stance*. Cambridge: MIT Press/A Bradford Book.
Des Chene, D. 2001. *Spirits and clocks. Organism and machine in descartes*. Ithaca: Cornell University Press.
Des Chene, D. 2005. Mechanisms of life in the seventeenth century. *Studies in the History and Philosophy of Biological and Biomedical Sciences* 36: 245–260.
Dowe, P. 2000. *Physical causation*. New York: Cambridge University Press.
Garson, J. 2008. Function and teleology. In *Companion to philosophy of biology*, ed. A. Plutynski and S. Sarkar. Malden: Blackwell.
Gettins, P.G. 2002. Serpin structure, mechanism, and function. *Chemical Reviews* 102: 4751–4804.
Glennan, Stuart S. 1996. Mechanisms and the nature of causation. *Erkenntnis* 44: 49–71.
Glennan, Stuart S. 2002. Rethinking mechanistic explanation. *Philosophy of Science* 69(Supplement): S342–S353.
Glennan, Stuart S. 2009. Productivity, relevance and natural selection. *Biology and Philosophy* 24(3): 325–339.
Good, B.L., and M.J. Eck. 2007. Mechanism and function of formins in the control of actin assembly. *Annual Review of Biochemistry* 76: 593–627.
Gould, S.J., and R.C. Lewontin. 1978. The spandrels of San Marco and the panglossian paradigm: A critique of the adaptationist programme. *Proceedings of the Royal Society of London* 205: 581–598.
Gould, S.J., and E. Vrba. 1982. Exaptation – A missing term in the science of form. *Paleobiology* 8: 4–15.
Haugeland, J. 1998. *Having thought*. Cambridge, MA: Harvard University Press.
Hempel, C.G. 1965. *Aspects of scientific explanation and other essays in the philosophy of science*. New York: Free Press.
Hull, D. 1974. *Philosophy of biological science*. Englewood Cliffs: Prentice-Hall.
Kauer, J.A., and R.C. Malenka. 2007. Synaptic plasticity and addiction. *Nature Reviews. Neuroscience* 8: 844–858.
Kauffman, S.A. 1971. Articulation of parts explanation in biology and the rational search for them. In *PSA 1970*, ed. R.C. Buck and R.S. Cohen. Dordrecht: Reidel.
Leunissen, M. 2007. The structure of teleological explanations in Aristotle: Theory and practice. *Oxford Studies in Ancient Philosophy* 33: 145–178.
Lycan, W. 1987. *Consciousness*. Cambridge, MA: Bradford Books/MIT Press.
Machamer, P.K. 1977. Teleology and selective processes. In *Logic, laws, and life: Some philosophical complications*, Pittsburgh series in philosophy of science, ed. R. Colodny, 129–142. Pittsburgh: University of Pittsburgh Press.
Machamer, P.K., L. Darden, and C.F. Craver. 2000. Thinking about mechanisms. *Philosophy of Science* 67: 1–25.
Mackie, J.L. 1977. *Ethics: Inventing right and wrong*. Pelican Books, Middelessex.
Mayer, et al. 1991. NMDA receptors: Physiological studies with divalent cations and competitive antagonists. In *Neuroscience of the NMDA receptor*, ed. A. Kosikowski and G. Barinuevo, 15–35. New York: VCH Publishers.
Millikan, R. 1984. *Language, thought, and other biological categories: New foundations for realism*. Cambridge, MA: MIT Press.
Neander, K. 1991. Functions as selected effects: The conceptual analysts defense. *Philosophy of Science* 58: 168–184.
Osler, M. 2001. Whose ends? Teleology in early modern natural philosophy. *Osiris* 16: 151–168.
Rowan, M.J., I. Klyubin, W.K. Cullen, and R. Anwyl. 2003. Synaptic plasticity in animal models of early Alzheimer's disease. *Philosophical Transactions of the Royal Society of London. Series B, Biological Sciences* 358(1432): 821–828.
Ruse, M. 1971. Functional statements in biology. *Philosophy of Science* 38: 87–95.
Russell, G. 2010. In defense of Hume's law. In *Hume, 'Is' and 'Ought': New essays*, ed. C. Pidgen. New York: Palgrave-MacMillan.
Salmon, W.C. 1984. *Scientific explanation and the causal structure of the world*. Princeton: Princeton University Press.

Salmon, W.C. 1989. Four decades of scientific explanation. In *Scientific explanation, Minnesota studies in the philosophy of science XVIII*, ed. P. Kitcher and W. Salmon, 3–219. Minneapolis: University of Minnesota Press.

Salmon, W.C. 1994. Causality without counterfactuals. *Philosophy of Science* 61: 297–312.

Schaffner, K.F. 1993. *Discovery and explanation in biology and medicine*. Chicago: University of Chicago Press.

Sellars, W. 2007. *The space of reasons: Selected essays of Wilfrid Sellars*, ed. R. Brandom and K. Scharp. Cambridge MA: Harvard University Press.

Shapin, S. 1996. *The scientific revolution*. Chicago: University of Chicago Press.

Shepherd, G. 1994. *Neurobiology*, 3rd ed. New York: Oxford University Press.

Sober, E. 1984. *Conceptual issues in evolutionary biology*. Cambridge, MA: Bradford/MIT Press.

Thagard, P. 2000. Explaining disease: Correlations, causes and mechanisms. In *Explanation and cognition*, ed. R. Keil and R. Wilson. Cambridge, MA: MIT Press.

Toulmin, S. 1975. Concepts of function and mechanism in medicine and medical science. In *Evaluation and explanation in the biomedical sciences*, ed. H. Tristram Engelhardt Jr. and S.F. Spiker, 51–66. Dordrecht: Reidel.

Westfall, R.S. 1971. *The construction of modern science*. New York: Wiley.

Westfall, R.S. 1973. *Science and religion in seventeenth-century England*. Ann Arbor: University of Michigan Press.

Wimsatt, W.C. 1972. Teleology and the logical structure of function statements. *Studies in History and Philosophy of Science* 3: 1–80.

Wimsatt, W. 1974. Complexity and organization. In *PSA 1972*, Boston studies in the philosophy of science, vol. 2, ed. K.F. Schaffner and R.S. Cohen, 67–86. Dordrecht: Reidel.

Wimsatt, W.C. 1976. Reductive explanation: A functional account. In Sober 1984, *Conceptual issues in evolutionary biology*, 477–508. Cambridge, MA: Bradford/MIT Press..

Woodward, J. 2003. *Making things happen*. New York: Oxford University Press.

Wright, L. 1973. Functions. *Philosophical Review* 85: 70–86. Repr. In Sober 1984, 347–368.

Zhao, C., W. Deng, and F.H. Gage. 2008. Mechanisms and functional implications of adult neurogenesis. *Cell* 132: 645–660.

Understanding the Sciences Through the Fog of "Functionalism(s)"

Carl Gillett

Abstract Versions of "functionalism," and their frameworks, have come to dominate many philosophical debates. Unfortunately, there is now a damaging interpretive "fog" where "functionalism(s)" is applied because it widely assumed that what I term the Standard Picture is true of the metaphysics of "functionalism" and hence that there are unitary notions of "functional property," "causal role," and "realization" based around the machinery of topic-neutral Ramseyfication and second-order properties. In this chapter, I use the case of the special sciences, and a version of "functionalism" based upon them, to show that the Standard Picture is deeply flawed. I show that the functional properties found in mechanistic explanations in the special sciences, as well as the versions of "functionalism" built upon them, fail to fit under the Standard Picture. I also highlight the flawed arguments about special sciences that have recently been driven by using the Standard Picture. In concluding, however, I outline some general meta-methodological lessons that can finally help to lift the "fog" enveloping "functionalism(s)" to ground more productive approaches in future work.

The 1960s, 1970s, and 1980s were a seminal period for the new "naturalistic" approach in philosophy. In particular, the engine for much of this work was the philosophy of mind and the philosophy of psychology which came into shape as we now know them during this time. Taking on foundational issues in these areas, as well as questions spanning philosophy of science, "naturalistic" metaphysics, and more, the founding generation of writers like Hilary Putnam, Jerry Fodor, and Daniel Dennett looked to the sciences to map out a range of philosophical frameworks and positions.[1] In particular, these writers defended a view of special sciences,

[1] Dennett (1969, 1978), Fodor (1968a, b, 1974) and Putnam (1960, 1973).

C. Gillett (✉)
Department of Philosophy, Northern Illinois University, DeKalb, IL, USA
e-mail: cgillett@niu.edu

P. Huneman (ed.), *Functions: selection and mechanisms*, Synthese Library, Studies in Epistemology, Logic, Methodology, and Philosophy of Science 363,
DOI 10.1007/978-94-007-5304-4_9, © Springer Science+Business Media Dordrecht 2013

and their properties, that I term "Mechanistic Functionalism" which sought to articulate the "functional properties," "causal roles," and "realization" found in mechanistic explanations in these disciplines.[2]

Mechanistic Functionalism, and its picture of the special sciences, famously played a central role in turning back the positivist's jaundiced view of these disciplines. But a range of other versions of "functionalism" were also developed for different purposes in this welter of early creative work. Some varieties were also empirically inspired, though focused more narrowly, such as forms of "functionalism" using as a basis the concepts found in computational approaches in cognitive science. Still other versions of "functionalism" were developed by philosophers, such as David Lewis, deploying more traditional, analytic techniques.

The importance of versions of "functionalism" cannot be overstated since they went on to play central roles not just in debates over the special sciences or the possibility of scientific reduction but also in wider discussions over the formulation of physicalism, the nature of mental causation, the character of the mind-body problem, and much more. Given this important role, it is especially unfortunate that there is now an ingrained tradition of referring to "functionalism" as a unitary position, with unitary concepts of "functional property," "causal role," and "realization," despite the actual variety of positions, frameworks, and concepts. In particular, though there may be many other controversies about "functionalism," there is a surprising consensus about the foundational question of the metaphysics of "functionalism" in what I term the "Standard Picture" – a position largely built around the technical machinery developed by analytic philosophers, primarily Lewis, and the concepts that grew from it (Lewis 1966, 1972, 1994).

The claims of the Standard Picture are familiar from textbooks, online encyclopedias, and graduate seminars.[3] First, it is assumed that the individuative features of "functional properties" are capturable by the machinery of topic-neutral Ramseyfication. Second, and largely growing out of the latter point, "functional properties" are taken to be what are termed "second-order properties" – that is, they are taken to be the properties of having some "realizer" property that plays a certain "causal role." Third, and building upon these background assumptions, the Standard Picture claims there are just *two metaphysical varieties* of functionalism: "Realizer Functionalism" under which "functional properties," taken to be second-order properties, are identical to the lower level "realizer" properties that play their causal roles, or "Role Functionalism" which takes "functional properties" to be second-order properties whose "roles" may be played by a variety of differing realizer properties to none of which such properties are identical.

Given the variety of versions of "functionalism" developed in the earlier period, one can only be suspicious that serious difficulties would plausibly arise when these

[2] This form of functionalism has been independently highlighted as a neglected position by both Gillett (2007a) and Piccinini (2010). The apt term "Mechanistic Functionalism" is from Piccinini.

[3] See, for example, Block (1980a, 1994), Kim (1996), Rey (1997), and Levin (2006), among many others.

different forms of "functionalism" are all shoehorned into the framework of the Standard Picture. Elsewhere I have documented just such difficulties by distinguishing the distinct proprietary concepts of some of the different forms of "functionalism."[4] However, in this chapter, I take a more focused approach to highlight these problems and look solely at the case of the special sciences and their properties, as well as Mechanistic Functionalism based upon them. My narrow goal is to show that when the Standard Picture is applied to either the notions of the special sciences or Mechanistic Functionalism, then it mischaracterizes them – with a host of consequent problems in the shape of flawed interpretations and arguments. Using these results, however, I then also highlight the meta-methodological lessons that we must learn in order to lift the damaging interpretive "fog" now generally enveloping discussions of "functionalism(s)."

I begin this chapter, in Sect. 1, by examining a concrete case from the special sciences. I detail both the *intra*-level role of mechanistic explanations as well as the *inter*-level mechanistic explanations that build upon them to more carefully draw out the notions of "functional property," "causal role," and "realization" implicit in such scientific explanations.[5] I also detail how these notions underpin Mechanistic Functionalism. I then remind the reader about the claims of the Standard Picture, in Sect. 2, focusing especially upon giving detailed accounts of its proprietary notions of a "functional property," "causal role," and "realization." (Throughout this chapter, I therefore leave explicit notions of "function" to one side since the trinity of concepts of "functional property," "causal role," and "realization" are more central to all versions of "functionalism.")[6]

I bring these two bodies of work together, in Sect. 3, to examine whether special science concepts, and the notions of Mechanistic Functionalism, fit under the Standard Picture. I offer a range of reasons why they do not. These detailed arguments

---

[4] See Gillett (2007a).

[5] Throughout this chapter, I therefore focus on the "functional properties" posited in mechanistic explanations in the special sciences since looking at these properties suffices to show that the Standard Picture fails to cover key concepts of the special sciences and Mechanistic Functionalism. I should not therefore be taken to be claiming that such notions are the only concepts of a "functional property" deployed in the sciences. (See note 5 below.)

[6] The reader should carefully note that I intend to take no stance in the recent debates over the kinds of "function" posited in the sciences. There is now a very mature literature on the nature of the "functions" found in the biological and other sciences. (For instance, see the papers in Bekoff et al. (1998), Buller (1999), and many of the other papers in this volume.) My account of the "functional properties" found in mechanistic explanations is compatible with other kinds of "function" and "functional property" than causal ones being used in other scientific areas. I am simply committed to a certain kind of "functional property" being deployed in mechanistic explanation, but neutral over the further kinds of "functional property," or "function," used in other scientific areas. As an aside, however, I should note that many of these further notions deployed in the sciences either build upon, grow out of, or are in some way connected to such causal notions. See, for example, Amundsen and Lauder (1998) and Craver (2012) for arguments in defense of this kind of point and more discussions of the relations of the concepts utilized in mechanistic explanations and in other types of explanation.

provide substantive grounds for thinking that the Standard Picture mischaracterizes the features of special sciences and Mechanistic Functionalism. Against this background, I detail a number of prominent cases, in Sect. 4, where the interpretive "fog" resulting from the Standard Picture has led to problems in a number of debates over the special sciences.

My main conclusions are that there are very real dangers in trying to understand the special sciences using the Standard Picture of "functionalism," including the machinery of topic-neutral Ramseyfication or notions like that of a second-order property. In contrast, I show that Mechanistic Functionalism provides a sympathetic framework for many phenomena in the special sciences. And, in concluding this chapter, I end on a still wider positive note by sketching three general, meta-methodological lessons that, if followed, allow us to leave the present "fog" surrounding "functionalism(s)" to pursue more productive approaches in future work.

1 Special Sciences and Mechanistic Functionalism: Functional Properties, Causal Roles, and Realization in the Sciences

A number of early proponents of "functionalism" like Jerry Fodor (1968a, b), Daniel Dennett (1969, 1978), Robert Cummins (1975, 1983), and William Lycan (1987, 1994) all focused on what were termed "functional" or "mechanistic" explanations (or "analyses") in the special sciences and the "functional" properties posited in them. (Throughout this chapter by "functional properties of the special sciences," I thus mean the properties posited in mechanistic explanations.) The resulting view is Mechanistic Functionalism which is a *general* account of the special sciences that utilizes the work on mechanistic explanation to ground its notions of "functional property," "causal role," and "realization." Building on this position, a number of these writers also argued that the emerging psychological sciences and their entities were *continuous* in nature with such special sciences and their entities, thus concluding that psychology was an equally legitimate science studying the same kind of "functional properties." The resulting view is a subspecies of Mechanistic Functionalism I have elsewhere termed "Continuity Functionalism."[7]

My primary focus here is not to give an exegesis of Fodor, Dennett, or Lycan's historical views, though I contend our work can be used to give such a charitable reconstruction. Instead, I intend to illuminate both the concepts of the special sciences, and also an "idealized" form of Mechanistic Functionalism, to use in the philosophy of science and in our assessment of the Standard Picture. To start, I am going to examine a concrete scientific example to illuminate what Fodor (1968b) terms the two "phases" of mechanistic explanation in the special sciences.[8] I therefore

---

[7] For earlier, and later, defenses of such claims, see Fodor (Fodor 1968a, b, 1994).

[8] Fodor (1968b), Chapter 3.

need a metaphysical framework to guide my work. Since our primary focus will be on properties, I follow debates in the metaphysics of science and assume a weakened version of the causal theory of properties under which a property is individuated by the causal powers it potentially contributes in this world, under certain conditions, to the individuals in which it is instantiated. As we will see, the causal theory of properties provides an especially congenial and illuminating framework for the concepts of the special sciences.[9]

As our example, consider a prominent recent case from the neurosciences. We know that, under appropriate background conditions, a potassium ion channel plays a key role in a neuron due to its property of being a voltage-sensitive gate contributing the backward-looking power of opening in response to a change in the charge of surrounding cells. As a result, we can appreciate what Fodor (1968b) called "phase one" of work in special sciences in what he terms "functional analyses" and which I will term an "*intra-level*" mechanistic explanation of why the ion channel opens in terms of the mechanism in which this individual is involved. For when the ion channel is under certain background conditions, such as being in the oily membrane of the cell, and there is a change in the charge of surrounding cells, then the backward-looking power, contributed by the property of being a voltage-sensitive gate, of the ion channel is manifested and grounds a causal process, or "mechanism," that results in the ion channel opening. Thus, the ion channel's property of being a voltage-sensitive gate, and the powers it contributes, allows us to explain the opening of the ion channel in the relevant conditions – an intra-level explanation of entities using other entities at the same level.

Note what such intra-level mechanistic explanations implicitly assume about the "functional properties" posited in them such as the property of being a voltage-sensitive gate. First, we can see that such properties are, in a sense, individuated by "causal roles." For such properties are taken to be active causes as causal *role-players* – they cause, and/or are caused by, specific properties at their own levels, through the forward- and backward-looking powers they contribute to individuals. For the powers of such properties underpin our intra-level mechanistic explanations. Second, such a functional property is individuated by what I term an "M-role," or "causal-mechanist role." An M-role refers to what individuals do, in causing (or being caused by) entities at their level, in virtue of the powers the relevant special science property contributes to them. "Functional properties" in the special sciences are thus plausibly taken to be "first-order" properties, for the same reasons most efficacious, causal role-playing properties appear to be "first order" – they are the properties taken to causally play their own defining roles. Third, we should mark that the individuating features of such properties are *specific* causal relations to other special science properties at the same level and hence by what I term "specific" causal roles. Thus, the property of being a voltage-sensitive gate is caused by the

[9] This framework is thus a variant of Shoemaker (1980). In addition, I will use "entity" in the standard way as a catchall referring to powers, properties, individuals, processes, etc.

property of surrounding charge changing, and causes other particular properties, such as having certain flows or concentrations of ions.

As I hope is now clear, but is worth emphasizing for our later discussion, special science properties have M-roles which these properties obviously play themselves, and such M-roles are specific causal roles individuated by their causal relations to specific properties of individuals at the same level. Special science properties are thus "functional," in the sense relevant to mechanistic explanations, because they are individuated by what they causally *do*, or what they causally result in if you prefer, and not by what they are "composed by." Pursuing the first intra-level phase in some scientific area, we come to have better and better intra-level mechanistic explanations. And, given the latter points, it is perhaps unsurprising that such intra-level explanations can be developed even though we presently know little, or anything, about what lower level properties, if any, the relevant properties depend upon. We should thus mark that very often in individuating special science properties we usually do not need to quantify over other properties. And we posit such properties regardless of whether we know they have other properties upon which they depend. I use the term "M-functional property," or "causal-mechanist functional property," to refer to properties of this type which are individuated by their M-roles and hence their causal powers.[10]

However, special science research does usually move on to what Fodor (1968b) terms "phase two" where researchers offer what Fodor dubs "mechanistic analyses" and what I term "*inter-level*" mechanistic explanations. Such inter-level mechanistic explanations allow us to understand the properties and other entities posited in intra-level explanations by illuminating lower level entities that compose them. For example, in the case at hand, Roderick MacKinnon won the Nobel Prize for his work establishing a compelling inter-level mechanistic explanation of how such ion channels open by illuminating the chemical and spatial properties/relations of the complex protein molecules that are "subunits," that is, parts, of these channels. Basically, as Fig. 1 illustrates, when the charge in surrounding cells changes, then the backward-looking powers, to change relative spatial position, of each of the subunits are manifested *together,* and the subunits all swivel to adopt new spatial relations to each other. These lower level mechanisms together implement the higher level process of the ion channel opening. And the many properties and relations of the subunits, such as their alignment and chemical properties, *together* realize the ion channel's property of being a voltage-sensitive gate – basically, the powers contributed by the lower level properties *together noncausally result* in the qualitatively different powers of the realized property.

[10] The reader may wonder whether this means that very many scientific properties, whether chemical, biological, neurophysiological, etc., are M-functional properties. This is indeed the case, so one might question what the difference between "structural" and "M-functional" properties actually is. In fact, the writers focused on the special sciences, and notions of "functional property" built upon their concepts, have long pressed the point that properties at all scientific levels are "functional." Thus, Lycan (1987) notes "functional-structural" designations are relative. And these points are backed by Shoemaker (2007) who has recently pressed the point that most properties are "functional" and that the "functional-structural" distinction is basically one concerning kinds of concepts, rather than properties.

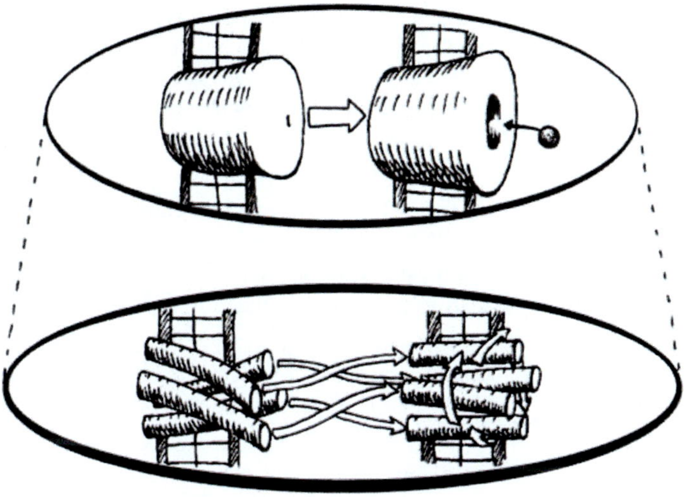

Fig. 1 A diagram of the lower level mechanisms at the *bottom*, involving the protein subunits, that implement the process of the potassium ion channel opening in response to a change in charge of nearby cells, outlined in the *top* of the figure

In such inter-level explanations we have an array of compositional relations holding between powers, properties, individuals, and mechanisms at higher and lower levels. But it is worth noting that the compositional relations that underpin inter-level mechanistic explanations all have a number of common features.[11] Let us consider just three of these features here. First, we should note that such relations, whether realization between properties or implementation between processes, are transitive, asymmetric, ontological determination relations which are what I term "noncausal" – such relations are synchronous, occur between entities that are not wholly distinct, and do not involve the mediation of force and/or the transfer of energy.

Second, we must carefully mark that scientific composition usually has *qualitatively different* relata – for the entities at distinct levels are different in their features. Thus, the properties and relations of the protein subunits contribute no common powers with the property of the ion channel – the protein subunits have powers such as changing their relative spatial positions, under certain conditions, but not the powers of opening or allowing speedy passage to potassium ions. Furthermore, these higher and lower level properties are also obviously instantiated in different individuals – in the ion channel and subunits.

Third, we can understand this puzzling feature by noting that although each lower level component entity is qualitatively distinct from the composed entity, the numerous lower level components *together* noncausally result in the qualitatively

[11] For more detailed accounts of the particular compositional relations between powers, properties, individuals, and processes in inter-level explanations, see Gillett (2007b, unpublished).

different composed entity. Thus, although the properties/relations of the subunits share no common powers with the property of being a voltage-sensitive gate, nonetheless *together* the contributions of powers by the properties/relations of the subunits can noncausally result in the powers individuative of this very different property.

Using these more general points, we can understand the so-called realization that holds between properties at different levels and which underpins inter-level mechanistic explanations. This is what I have elsewhere termed "causal-mechanist" or "M-realization" of the "Dimensioned" variety (Gillett 2002). We can offer this simple "thumbnail" account of such Dimensioned M-realization as follows:[12]

> (Dimensioned M-Realization) Property/relation instance(s) F_1-F_n realize an instance of a property G, in an individual *s* under condition $, *if and only if s* has powers that are individuative of an instance of G in virtue of the powers contributed, under $, by F_1-F_n to *s* or *s*'s constituent(s), but not vice versa.

The Dimensioned account nicely covers key features of scientific realization relations. For it earns its name because we have "dimensions" in the form of distinct lower and higher level powers, properties, and individuals, as well as the distinct processes they ground. Thus, among other features, the Dimensioned view allows for both the qualitative distinctness of realizer and realized properties and their being instantiated in distinct individuals, and also for the many-one character of scientific realization.

As an aside, the reader should briefly note how the general features of Dimensioned M-realization allow for the so-called "multiple" realization used by Putnam and Fodor against the positivists. The relata of Dimensioned M-realization are instances of qualitatively different properties sharing no common powers, and so very different combinations of lower level properties may each result in the same powers in those individuating the higher level property. Consequently, we have the possibility of *multiple* lower properties whose instances each M-realize instances of the *same* higher level property.[13] Famously, Putnam and Fodor argued that in many actual cases we have just such a situation, hence blocking the inter-level property identities demanded by the positivists' Nagelian model of reduction.

To summarize, we have learned a lot about special science concepts from a brief examination of both of the main "phases" of mechanistic explanation. Given their roles in intra-level mechanisms and mechanistic explanations, we saw that special science properties are M-functional properties in the sense of being causal role-players individuated by the causal mechanisms that the manifestation of their powers grounds, rather than being individuated by the entities that realize them. Special science properties are thus individuated by M-roles that refer to what individuals do,

---

[12] Elsewhere, I offer a full account of such Dimensioned causal-mechanist realization as part of an integrated view of the compositional relations between powers, properties, individuals, and mechanisms (Gillett unpublished).

[13] See Aizawa and Gillett (2009a, b) for a detailed account of such multiple realization in the sciences.

in causing (or being caused by) entities at their level, in virtue of the powers these properties contribute to such individuals. Consequently, we also found that there is often *nothing compositionally in common* to all of the instances of such functional properties when they are multiply M-realized, so such properties are instead naturally thought of as "functionally" individuated by the causal-mechanist roles which they all play. Finally, such special science properties may, or may not, have been shown to be composed by lower level scientific properties. But when we have such "realization," it takes the form of the ontological M-realization we have labeled "Dimensioned" since it often involves qualitatively different powers, properties, individuals, and processes at different levels.

Having illuminated the notions of "functional property," "causal role," and "realization" we find in mechanistic explanations in the special sciences, we also have a better understanding of the Mechanistic Functionalism that Fodor, Dennett, or Lycan built upon these concepts. In a "pure" or "idealized" form, untainted by some of the later adulterations I note below, the Mechanistic Functionalism pioneered by such early writers is focused on M-functional properties individuated by the specific causal-mechanist roles these properties themselves causally play, in virtue of the powers these properties contribute to individuals, and often, though not necessarily, M-realized in the Dimensioned manner by lower level properties – and often multiply M-realized by such properties.

2 The Fog Descends?

2.1 The "Standard Picture" of the Metaphysics of Functionalism

Although earlier writers, like Fodor or Dennett, defended versions of "functionalism" guided by the features of mechanistic explanations, the idealized version of Mechanistic Functionalism outlined in the last section is now barely recognizable to many philosophers as a form of "functionalism." The reasons for this change over the intervening decades are numerous.[14] One is that, as its original proponents now acknowledge, early versions of Mechanistic Functionalism were altered to incorporate concepts drawn from computational work in cognitive science. For example, Fodor has recently explained that:

> ...getting clear on the nature of the project took considerable time and effort. Particularly striking in retrospect was the widespread failure to distinguish the computational program in psychology from the functionalist program in metaphysics; the latter being, approximately, the idea that mental properties have functional essences... (For an instance, where the two are run together, see Fodor 1968a.) (Fodor 2000, p.105, Fn. 4).

[14] For a careful examination of some of these historical issues, see Piccinini (2004).

The notions of "functional property," "causal role," and "realization" drawn from computational theories thus led to revised forms of "functionalism" growing out of, but ultimately rather different from, Mechanistic Functionalism.[15] However, the most important factor in the alterations to early forms of Mechanistic Functionalism, often reinforcing the first set of changes, was the eventual dominance of the theoretical machinery developed by analytic philosophers for their versions of "functionalism."

Unfortunately, the empirically oriented proponents of early versions of "functionalism" did little to more precisely articulate the notions of "functional property," "causal role," and "realization" that they used in their Mechanistic Functionalism, perhaps thinking that such concepts were clear in the scientific cases to which they pointed. But the result was a theoretical vacuum, and this was ultimately filled by Lewis's machinery of topic-neutral Ramseyfication, and the notions that grew from it, to the point where this framework is now the dominant account of the metaphysics of "functionalism" in what I am terming the "Standard Picture." Once again, as this machinery was used to interpret all versions of "functionalism," the result was important alterations to Mechanistic Functionalism – or so I shall argue in Sect. 3.

Given its importance, let us remind ourselves about the nature of the Standard Picture which has three core claims we may frame thus:

> (The Standard Picture) A "functional property" (i) has its individuative features characterized by topic-neutral Ramseyfication and a generic causal role, and (ii) is a second-order property, that is, the property of having a realizer property that plays a certain role. As a result, (iii) there are two metaphysical kinds of functionalism: either one endorses "Realizer Functionalism" which takes functional properties to be identical to their realizers or one accepts "Role Functionalism" and takes functional properties to be realized by a variety of different realizers.

Though the Standard Picture is familiar, I want to more carefully probe its notions. First, I detail the role of topic-neutral Ramseyfication and, most importantly, the idea of a "functional property" as a second-order property which it spawned. In addition, I detail the differing notions of a "causal role" and "realization" that go along with the Standard Picture, and I conclude the section by sketching Realizer and Role functionalism.

Let us begin with a quick outline of the machinery of topic-neutral Ramseyfication.[16] To this end, consider T which is a theory, or some other set of sentences,

---

[15] There is insufficient space to look at the differing notions of "casual role," "functional property," and "realization" inspired by computational approaches, so let me just focus on "realization" to highlight these divergences. There is a kind of computational or mathematical relation commonly referred to as "realization," in mathematics, the sciences, and philosophy, which I term "Abstract," or "A-realization." Very crudely, X is taken to A-realize Y if the elements of X map onto, or are isomorphic with, the elements of Y. This notion of "realization" is commonly utilized with formal or computational models, and we should note that the relata of such A-realization relations are largely unconstrained and need not, like M-realization, have causally individuated entities as relata. For A-realization holds simply in virtue of a bare abstract mapping which can obviously hold between all manner of entities.

[16] Here, I loosely follow the presentation of Block (1994).

about the psychological property (or state) of pain and which has n mental terms of which the 17th term is "pain." Furthermore, assume the relevant mental properties (or states) discussed by T are characterized by their relations to n inputs and n outputs. Using topic-neutral Ramseyfication on the sentences of T, we can then define "pain" relative to T, where F1-Fn are variables that replace the n mental terms of T, *i*1-*i*n refer to the relevant inputs (such as bodily damage), and *o*1-*o*n refer to the relevant outputs (such as wincing). The resulting definition of pain is the following

Being in pain = Being an x such that ∃F1...∃Fn[T(F1...Fn,*i*1...*i*n,*o*1...*o*n,etc) & x is in F17]

Here on the right we have a topic-neutral Ramsey sentence that is taken to define the predicate on the left. Furthermore, as we shall see below, such Ramsey sentences are now also taken to individuate the properties to which such predicates refer: thus, the Ramsey sentence is taken to articulate the "causal role" played by the property referred to by the relevant predicate. For obvious reasons, that will become even clearer shortly, I term these "L-" or "linguistic roles" since they are sentences. In addition, we should also note that the machinery of Ramseyfication also supplies a characteristic notion of "realization." As should be clear, the machinery of topic-neutral Ramseyfication was intended by Lewis to apply in the "formal" mode to predicates and other semantic entities. It thus comes as no surprise that under this view, F is realized by some physical property P just in case P satisfies the topic-neutral Ramsey sentence for the predicate "F." I will refer to this semantic relation, whose relata are a property and a sentence, as "L-" or "linguistic" realization.

The machinery of topic-neutral Ramseyfication stands at the heart of the Standard Picture, so it is important to note its topic-neutrality. Lewis was heavily influenced by the work of the identity theories of U.T. Place and especially J.J.C. Smart. As a result, Lewis's machinery was made to be "topic-neutral" – that is, to exclude all explicit use of mental predicates or properties when characterizing a mental predicate or property. This topic-neutrality followed a tactic of Smart who was keen to ameliorate what he saw as "suspect" or "illegitimate" nature of mental properties. The idea was that one could not characterize mental properties in terms of their relations to other "suspect" mental properties. In response to this "problem," topic-neutral Ramseyfication famously characterizes properties, whether mental or otherwise, in terms of what I shall dub "generic" causal roles. For using topic-neutral Ramseyfication, properties are taken to be characterized by, that is, individuated by, causal roles of causing *some* property that causes *some* property that causes *some* property, in a complex web of causal relations between topic-neutrally characterized properties. We consequently get thesis (i) of the Standard Picture which takes "functional" properties to be characterized using topic-neutral Ramseyfication and hence to be individuated by generic causal roles.

Given the nature of topic-neutral Ramseyfication, we can also appreciate how it led to the key notion of a "second-order property." For example, consider this telling passage from an early paper on functionalism by a proponent of the Standard Picture. Robert Van Gulick tells us:

> The psychological property F is not to be identified with the set of properties which are related in way R, but rather with the property of having properties which are related in way R. Psychological properties, involving as they do quantifications over properties are second order properties. Thus it is argued that they cannot be identified with any straightforwardly physiological properties which are correctly understood as first order structural properties. (Van Gulick 1983, p. 188).

This passage is characteristic of a number of features involved with the Standard Picture, but the point I want to focus on is the way that the use of topic-neutral Ramseyfication configures Van Gulick's deeper assumptions about a "functional property," in this case a psychological property.

It should immediately be striking that a philosopher as adept as Van Gulick is led into what appears to be an uncharacteristic category mistake for quantification is obviously a semantic or logical operation, and one may wonder how an ontological entity such as a property can "involve" quantification. The obvious explanation of why Van Gulick says something so odd is that he accepts that topic-neutral Ramseyfication tells us what individuates a "functional property" when a Ramsey sentence is taken "materially" as individuating a property, rather than just a predicate. For if the topic-neutral Ramsey sentence captures what a "functional property" is, then such a property would appear to be the property of having some property that plays the generic causal role laid out in the Ramsey sentence. If we term as "first order" the property that plays the role, what Van Gulick terms a "structural property," then it appears natural to say that the "functional property" is "second order" – for it is the property of having some first-order property that plays a certain "role." We are thus brought to thesis (ii) of the Standard Picture.

It is worth noting an important feature of a second-order property. For it appears highly plausible that a second-order property is a necessarily realized, or at least a necessarily dependent, property. Such properties always have to hold the hands of other properties in order to exist at all by their very definitions. For a second-order property is only ever instantiated in the world in virtue of some property playing its individuating role.[17] In a sense, second-order properties are rather sad properties that have never managed to leave "home" – second-order properties are always tied to their "mother" realizer property that must accompany them everywhere and play the roles of the sad second-order properties which are incapable of playing their own roles.

Topic-neutral Ramseyfication and second-order properties are common to all versions of the Standard Picture, but it is important to note that there are variations in this position. If one uses topic-neutral Ramseyfication in what we might term the "formal" mode of Lewis, then this machinery is focused on predicates and uses L-roles, in Ramsey sentences, and L-realization based around their satisfaction. However, though many do still use such notions in their understanding of the

[17] As well as Van Gulick (1983) and other primary sources, recent overviews of functionalism, such as Block (1994), Kim (1996), and Levin (2006), among many others, confirm the widespread understanding of "functional properties" as second-order properties.

Standard Picture, there has been a slow drift to what Ronald Endicott aptly terms a "material," ontological reading of the machinery of topic-neutral Ramseyfication focused on properties. Accompanying this evolution, there has been the development of distinct notions of "causal roles" and "realization" for such "material" versions of the Standard Picture.

For example, writers now often replace topic-neutral Ramsey sentences with causal-mechanist roles. However, these M-roles are heavily marked by the shadow of topic-neutral Ramseyfication. For the "functional properties" individuated by such M-roles are *still* taken to be second-order properties and hence are *still* taken to have a property that literally plays the very role of the second-order properties. This is important because it means that such "functional properties" are taken always to have *one* realizer property that goes along with them, shares the very same M-role and powers, and which is thus instantiated in the very same individual as the "functional property."

New notions of "realization" have also been developed in this "material" trend. Rather than L-realization, writers like Jaegwon Kim (1998), Sydney Shoemaker (2001), and Larry Shapiro (2004) have all endorsed what I have termed elsewhere the "Flat" view of M-realization as a component of the Standard Picture. And the marks of topic-neutral Ramseyfication can again clearly be seen in such Flat M-realization which is a one-one relation between a realizer and realized property, where the realizer property contributes all the powers of the realized property and hence plays the M-role of realized property. Unsurprisingly, Flat M-realization also takes the realized property and its realizer to be instantiated in the very same individual. Second-order properties defined by such M-roles, and engaging in Flat M-realization, are apparently taken to better fit the recent ontological, "material" readings of the Standard Picture.

We have now seen that the Standard Picture has its own proprietorial understanding of the trinity of key functionalist notions, taking a "functional property" to be a second-order property, "causal roles" to be either L-roles or M-roles, and "realization" to be either L-realization or Flat M-realization. The final element of the Picture is thesis (iii) and its claim that there are consequently just two kinds of metaphysical functionalism.

On one side, we have the Realizer Functionalism that is commonly attributed to Lewis. Realizer Functionalism is taken to use Arguments for Identity, grounded upon the machinery of the Standard Picture, to identify functional properties understood as second-order properties with their first-order realizer properties – hence the name Realizer Functionalism. The other metaphysical option is Role Functionalism that again takes topic-neutral Ramseyfication as its starting point, but is famously taken to argue, as Van Gulick did in our earlier quote, that a "functional property" is not identical to any realizer property. For it is noted that many different properties can L- or Flat M-realize the same second-order property. Many different properties can satisfy the relevant Ramsey sentence and hence L-realize the second-order property. Alternatively, taking "realization" to be Flat M-realization, many lower level realizers Flatly M-realize the same higher level property because these realizers contribute distinct sets of powers that all include the realized property's powers as a subset. Given this "multiple" L- or Flat M-realization, Leibniz's Law of Identity is claimed to fail for second-order properties and their realizers – thus, Role Functionalism is championed over Realizer Functionalism.

We have now examined the central notions, and theses, of the received wisdom about the metaphysics of "functionalism." And it is worth noting that, despite its widespread acceptance, prominent proponents of empirically oriented versions of "functionalism" have long been leery of the Standard Picture. Thus, in his careful introduction to the philosophy of mind, largely focused on the empirically oriented tradition, we find Georges Rey telling us in a footnote that:

> There are subtle issues that there is not space to pursue about how precisely to think of the metaphysics of functional *properties*. One standard way is to regard them as *second-order* properties:. i.e. properties of having things having some or other first-order (physical) properties that allow the role to be realized... I, myself, am dissatisfied with this approach (what do we say if the fundamental properties of physics turn out to be functional?), but have no better proposal... (Rey 1997, n.17, p. 183. Original emphasis)

Here we see an explicit concern about the key notion of a second-order property to which I return below. And such unease is implicitly mirrored in the work of writers like Fodor, Dennett, or Lycan who also usually eschew second-order properties and avoid using the machinery of topic-neutral Ramseyfication. It therefore appears empirically oriented "functionalists" have been suspicious of the Standard Picture, so in the next section I propose to examine whether these suspicions are well founded or not.

3 Special Science Properties, the Standard Picture, and Mechanistic Functionalism: Understanding the Problems

Our earlier work now allows us to examine whether "functionalism" as conceived by the Standard Picture provides a comfortable home for either the functional properties of the special sciences or Mechanistic Functionalism. I shall argue that it does not. First, I offer a number of reasons why the "functional properties" posited in the special sciences, and Mechanistic Functionalism, do not fit under the Standard Picture. I also note divergences in the notions of "realization" found in the special sciences from those embodied in the Standard Picture. I conclude the section by using our work to establish that Mechanistic Functionalism simply does not fit under either Realizer or Role Functionalism and hence that the Standard Picture embodies a *false dichotomy* about the metaphysical forms of functionalism. Overall, I therefore show that all the core theses (i)–(iii) of the Standard Picture are mistaken in the case of the special sciences and Mechanistic Functionalism.

The first problem is focused on the topic-neutrality of the Ramseyfication central to the Standard Picture. Once one considers real scientific cases, then we immediately confront problems with the use of topic-neutral Ramseyfication to capture the individuative features of special science properties. For example, the special science property of being a voltage-sensitive gate, and any other special science property, is individuated by what I earlier termed a "specific" causal role. Being a voltage-sensitive gate is individuated by a causal role which involves bringing about instances of particular special science properties, rather than the generic causal role that results from topic-neutral Ramseÿfication of causing *some* property, or other, that causes *some* property, or other, that causes *some* property or other. Even brief consideration of our earlier example shows that the sciences take the ion channel's property of being a voltage-sensitive gate to be characterized by its powers to cause, and be caused by, instances of specific special science properties – properties like having a certain charge or having a certain concentration or flow of ions. It is therefore plausible that special science properties are not identical to properties individuated by the generic causal roles outlined using topic-neutral Ramseyfication. And we can therefore see that thesis (i) is false for the "functional properties" posited in the special sciences and Mechanistic Functionalism.[18]

Though this conclusion runs counter to the received wisdom, if one steps back from recent philosophical debates, an obvious question immediately strikes one: *why* would it ever have been thought to be necessary to use topic-neutral Ramseyfication to characterize special science properties? Working scientists do not apparently use any such machinery, and the way that such scientists understand special science properties has made them spectacularly successful, at least in comparison with philosophers, in progressively illuminating the nature of the phenomena they study. One might thus plausibly conclude that the theories and explanations offered by the sciences do perfectly well in characterizing special science properties, leaving applications of the machinery of topic-neutral Ramseyfication unnecessary. As we noted earlier, this implicitly appears to have been the stance of prominent writers in the empirical tradition, like Fodor, Dennett, and Lycan, who have usually avoided using topic-neutral Ramseyfication in formulating their versions of "functionalism."

With these findings firmly in mind, we can also present a second array of problems for the Standard Picture centered upon thesis (ii) – that is, the claim that "functional properties," whether of the special sciences and/or Mechanistic Functionalism, are second-order properties. The first of these difficulties is that in the special sciences there is usually no "realizer" property that plays the very same M-role (or L-role) as that which individuates some higher level special science property. Instead, as we saw earlier, we usually have *many* lower level realizers, each playing a *different* M-role (and hence L-role) from the realized property, but where these qualitatively different realizers can *together* result in the realized property. But a second-order property is the property of having some property that plays the M-role (or L-role) associated with this property. If no such "realizer" properties

[18] This worry is obviously not a new one. For instance, David Chalmers (1996) presses the point that the 8ausal roles produced by topic-neutral Ramseyfication fail to properly characterize the individuating features of conscious experience for just the kind of reasons I have outlined. I have now argued that when we consider a wider set of disciplines than those of the psychological sciences, then we find that the similar points plausibly hold for the properties of all the special sciences, from chemistry to neurophysiology.

plausibly exist in the special sciences, then neither do such second-order properties. For instance, in the neurosciences we find no property at the molecular level that contributes powers like opening in response to a change in charge, or allowing speedy passage to ions, for the properties of the subunits only contribute powers such as changing their relative spatial positions, under certain conditions. Instead, as we noted earlier, we only have a group of many lower level properties none of which has the same M-role as the higher property, but whose powers together result in the powers of the higher level property. Consequently, we lack realizer properties that have the very same roles as the relevant realized property. And we can thus see that the functional properties of the special sciences cannot be second-order properties and that thesis (ii) is false since the kinds of realizer property required for second-order properties simply do not exist in the special sciences.[19]

These points are a little abstract and may leave the reader unconvinced, but they are reinforced by a second set of reasons to doubt that the functional properties of special sciences are second-order properties. This difficulty again concerns the fact that, by their very natures, second-order properties are essentially realized, or at least essentially dependent, properties – properties that must go around with their "mother" realizer properties that L-realize or Flatly M-realize them. In contrast, however, there are good reasons to think the functional properties of special sciences have no such feature of being essentially realized or dependent. (I leave to one side

[19] A common response to this objection focuses on so-called "structural" properties as "realizer" properties. In the scientific example at hand, the property of being a voltage-sensitive gate, instantiated in the ion channel, would have as its putative "realizer" the ion channel's putative structural property, call it "COMBO," of being "made of" protein subunits with certain spatial and chemical properties and relations, that is, the individuals, relations, and properties the sciences illuminate as being involved in M-realizing the ion channel's property of being a voltage-sensitive gate. This structural property at least has the chance to be the L-realizer or the Flat M-realizer of the ion channel's property of being a voltage-sensitive gate, for both properties are instantiated in the same individual and hence COMBO at least has the chance of playing the individuative causal role of the property of being a voltage-sensitive gate. (I put to one side our previous worry about whether the individuative causal role of the special science property is a generic or specific role since this will not affect my objection.)

Unfortunately, although structural properties may be an ideal fit for the demands of the Standard Picture, there is a grave concern that arises when structural properties, or similar entities, are used to understand scientific cases. For there are good reasons to think that we should not accept the existence of COMBO, and other structural properties, given the strong ontological parsimony arguments against positing any such properties. As we have seen in our scientific case, it is plausible that lower level property instances, putting it neutrally, "compose" or "make up" instances of higher level properties by together non-causally resulting in the powers individuative of the higher level property. The resulting picture is a compelling one which takes the "composing" or "making up" between properties, and other entities in the sciences, to be a determination *relation* – one holding between the properties and relations of the subunits and the properties of the ion channel. But, against this background, the introduction for the ion channel of a further *property of being made-up by* the subunits and their properties and relations, that is, a structural property like COMBO, looks almost perversely profligate.

the fact that functional properties in the special sciences play their own roles while sadly second-order properties must have other properties play their roles for them.)

As we have already noted, it is very often the case that we do not know whether a special science property is realized or by what. And historically, in the not so distant past, such a situation has obviously been extremely common for properties in all manner of special sciences. Given this state of affairs, it appears plausible that special science properties are not taken to be essentially realized properties, thus making it plausible that they are not second-order properties.

A different tack illuminates the same point and confirms Rey's suspicion that something is amiss. For even after we think we have evidence that a special science property is M-realized, it is always an open epistemic possibility for working scientists (however remote) that the property should still turn out to be, given new empirical discoveries, an ontologically fundamental entity which is neither realized by, or dependent upon, any other property. It is perfectly possible for any special science property *F* that some startling array of new empirical findings might undermine our past accounts of its composition and make it far more plausible that *F* is an ontologically fundamental property. But it appears that if this is an open epistemic possibility, then functional properties of the special sciences are not second-order properties – for it is not even an open epistemic possibility to discover that a second-order property is ontologically fundamental given that such a property is essentially dependent. One would instead discover there is no such property.

Such examples are not merely hypothetical and can be bolstered by evidence from actual scientific practice. Consider the famous case of the electromagnetic force. Initially, the electromagnetic force was a property posited in the special sciences and was not included in the fundamental ontology of the world, nor studied by fundamental physics. However, it became increasingly clear through further findings that, in fact, the electromagnetic force was plausibly an ontologically fundamental property and it was consequently included as one of the fundamental forces of nature, a status it continues to hold. We should note that if this special science property were a second-order property, then such a historical eventuality ought to involve our discovering there really is no electromagnetic force (the second-order property) and positing the existence of a new, first-order property. However, this is not what apparently occurred with the electromagnetic force. We continued to accept the very same property, but simply concluded it was an ontologically fundamental property – thus confirming our earlier point (and Rey's worry) that functional properties in the special sciences are usually not essentially dependent, or essentially realized, properties and hence not second-order properties. We can therefore see that there is a second reason to think thesis (ii) is false about the functional properties of the special sciences and Mechanistic Functionalism.

In addition to these concerns, our earlier work highlights problematic divergences between the "causal roles" and "realization" posited in the special sciences and under the Standard Picture. First, we saw that special science properties are commonly taken by working scientists to be *causal role-players* and hence to have an M-role, rather than to be individuated merely by an L-role and some sentence. Certainly, if a property is a causal role-player, then this usually means that associated with this property there will likely be some L-role in a Ramsey sentence derived from a predicate that describes the special science property. (Though note that undiscovered special science properties have M-roles despite lacking L-roles since no actual predicates have been coined to describe them.) However, the causal-mechanist role that individuates a property which is a causal role-player is not identical to an L-role, for the contribution of powers is clearly not identical to being described by a sentence. So we see a divergence about "causal roles" between the linguistically oriented versions of the Standard Picture and the special sciences.

More pervasive problems concern "realization." We again find problems with linguistic versions of the Standard Picture, for they posit L-realization for functional properties, while we have seen that the inter-level mechanistic explanations in the special sciences posit an ontological relation in a form of M-realization. And there are also difficulties with the more ontologically oriented, "material" interpretations of the Standard Picture that take functional properties to be Flatly M-realized – where "realization" is a one-one relation between a realizer and realized property, the realizer property contributes all the powers of the realized property, and this realizer is instantiated in the same individual as the realized property. The difficulty with Flat M-realization is that we saw in our scientific case that "realization" relations in the sciences are often many-one, with many lower level realizers that are qualitatively different from, and instantiated in distinct individuals than, the realized special science property. But Flat M-realization allows none of these features! Thus, although Flat M-realization may fit nicely with the topic-neutral Ramseyfication that philosophers have often been concerned with, it lacks the features of the realization relations we find in the special sciences. Instead, as we saw above, the functional properties of special sciences are usually Dimensionally M-realized by lower level properties.[20] We can thus see that the types of "causal role" and "realization" posited in the Standard Picture, whether those in early linguistic readings of it, or later "material" interpretations, simply do not fit the notions of "causal role" and "realization" we find in the special sciences and with Mechanistic Functionalism.

At this point, let us finally turn to thesis (iii). The question I want to consider is whether Mechanistic Functionalism falls under either Realizer or Role Functionalism, and hence under the Standard Picture. We can give a myriad of reasons why this is not the case, but let me focus on just a few.

It is clear that Mechanistic Functionalism is not a form of Realizer Functionalism. Recall that Mechanistic Functionalism takes higher level special science properties to be individuated by specific causal-mechanist roles that are qualitatively different from those of any other property. Mechanistic Functionalism thus takes these properties themselves to be first-order causal role-playing properties which do not share their causal-mechanist roles with, nor are identical to, any lower level properties. So Mechanistic Functionalism diverges from the Realizer Functionalism that defends identities between functional and realizer properties.

---

[20] See Gillett (2002) for an outline of these problems.

We can also quickly see that Mechanistic Functionalism, since it denies functional properties are second-order properties, does not fall under Role Functionalism either. Mechanistic Functionalism denies that any other property plays the M-role defining a functional property, and although it allows that functional properties in the special sciences may often be shown to be M-realized, this position denies that functional properties are necessarily realized by, or dependent upon, other properties. In contrast, Role Functionalism takes functional properties, as second-order properties, to be necessarily dependent properties which are always accompanied by realizers that play the M-role of the functional property.

We can thus see that Mechanistic Functionalism does not fall under either Realizer or Role Functionalism. So my final conclusion is consequently that the prevailing wisdom that there are only two metaphysical ways to be a functionalist, embodied in thesis (iii) of the Standard Picture, actually presents a *false dichotomy*. For Mechanistic Functionalism is clearly a version of "functionalism," but falls under neither of the metaphysical options presented by the Standard Picture. Rather than two ontological options, our work shows that there are *at least* three metaphysical variants of "functionalism." In fact, given the plethora of work on versions of "functionalism" we began this chapter by noting, offered in different areas for distinct ends, there are likely to still *more* metaphysical ways to be a functionalist than these three.

4 Accidents in the Fog: Some Misplaced Critiques

Our work has now shown, at some length, that the special sciences and their notions, as well as the Mechanistic Functionalism built upon them, simply do not fit under the framework and concepts of the Standard Picture. In many ways, when we have applied the Standard Picture to the work of empirically oriented "functionalists," like Fodor, Dennett, or Lycan, or used it directly to understand phenomena in the special sciences, then we have reason to think we have been in a damaging interpretive "fog." I want to briefly note some cases where I contend just this type of difficulty has arisen to show the reader that this situation has been far from benign.

Though there are many other instances, consider just these four recent, and very prominent, examples of such mistaken arguments:

(a) Objections that a "functionalism" about special sciences is not compatible with scientific realism (Pereboom 1991)
(b) The many arguments that "functional properties" in the special sciences cannot be causally efficacious (Kim 1998; Pereboom 2002)
(c) Critiques that seek to show "functionalism" about special science properties is inconsistent with multiply realized properties that are scientifically legitimate (Kim 1992; Shapiro 2000, 2004)
(d) Criticisms of accounts of "realization" in the special sciences that these accounts fail to fit the notion of "realization" of the Standard Picture (Polger 2007)

> I contend that in each of (a)–(d), among other cases, critics have assumed the Standard Picture, or elements of it, in making their various arguments. Unfortunately, such arguments simply do not go through if we use the notions of "functional property," "causal role," or "realization" utilized in the special sciences or under Mechanistic Functionalism.

For example, with regard to (a) or (b), anyone who understands the nature of a second-order property will be rightly dubious about whether we should be scientific realists about such properties or take them to be causally efficacious. However, we have seen that the functional properties of the special sciences, and also of the Mechanistic Functionalism of writers like the early Fodor or Dennett, are *not* second-order properties. Instead, functional properties in the special sciences are properties that contribute powers to individuals and play their own distinctive causal-mechanist roles. And scientific realism, or claims of causal efficacy, about such functional properties is far from *obviously* problematic.

Less obvious is how the Standard Picture might underpin (c) and recent metaphysical critiques of multiple realization, so let me say a little more about this case.[21] A feature of the Flat M-realization we find under the Standard Picture is that realizer and realized properties overlap in powers. Both Kim and Shapiro use this feature to argue that heterogeneity of realizers, and hence their powers, leads to heterogeneity of powers of realized properties in cases of multiple realization – and hence to conclude that multiply realized properties must be scientifically illegitimate properties that are not fit subjects of scientific laws because these properties vary in their powers.

Unfortunately, however, no such feature is found with the Dimensioned M-realization we have seen that we actually find in the special sciences and Mechanistic Functionalism. Dimensioned M-realization allows that realizers do not overlap in powers with realized properties since these lower level properties are usually qualitatively distinct from the higher level properties they realize. Consequently, heterogeneity of powers of realizers does not result in heterogeneity of powers of realized, *or multiply realized*, properties which all have the same powers and remain fit subjects for scientific laws. Once again, we see the damaging role of the Standard Picture which underlies a critique that simply does not go through under the actual concepts of the special sciences or Mechanistic Functionalism.

In a way, one can only feel sorry for such critics who have apparently acted in perfectly good faith in taking the Standard Picture to capture the metaphysical heart of all kinds of "functionalism." Nonetheless, though such critics may be innocent in their interpretations of "functionalism," the arguments that underlie (a)–(d) are still either question-begging or unsound. For, by focusing on the Standard Picture, and its notions, these critiques fail to establish that the relevant versions of "functionalism," or its proprietary concepts, actually lead to problems.

[21] See Gillett (2003) for more detailed versions of these arguments.

5 Conclusion – Meta-Methodological Ways to Dispel the Fog of "Functionalism(s)"

We have found strong reasons to believe that the recent use of the Standard Picture of "functionalism" to understand phenomena in the special sciences, or the work of empirically oriented "functionalists," has been dangerously flawed. Looking more widely, we can see that similar points plausibly apply generally to a range of debates and areas of philosophy. It is dangerous to simply use the Standard Picture, and its proprietary concepts of "functionalism," "functional property," "causal role," and "realization," to understand any phenomenon that might lend itself to understanding under a "functionalist" framework. It is equally dangerous to use the Standard Picture to understand any writer whose work is "functionalist" in nature. At this point in our discussions of "functionalism(s)," there is a desperate need for more meta-methodological care and reflection. In order for future work to be more productive, it is therefore important that we learn some relatively simple lessons.

First, we can plausibly see that we need to be more explicit and careful about what the goals of a research program are before we launch into action in assessing whether it succeeds. Research programs in the philosophy of science, the philosophy of mind, or even in biology, psychology, or mathematics all use common terms like "realization," and can be understood as committed to "functionalism" or "functional properties," but often express different concepts with these common predicates in the service of projects focused on different object phenomena. We need to be aware of, and respect, the diversity of research programs and the differences among them.

Second, we also need to keep clearly in mind that there is no single notion of "functionalism," "functional property," "causal role," "realization," and so on that is used by all research programs since these programs use distinct, and often proprietary, concepts. For, as we have seen, the Standard Picture is thus mistaken in many of its key claims. However, if one fails to heed this lesson, or the last, then one will unwittingly apply the proprietary concepts of a competitor in understanding some other research program – where the result is usually a weak and unstable position open to obvious objections. Unfortunately, as we saw in the last section, such criticisms of the relevant research program are usually unfounded since these worries often do not arise when the research program's own concepts are used in formulating its claims. We thus also need to be more careful about the diversity of concepts of "functionalism," "causal role," etc., and which of these concepts are most appropriately used in various cases, as we ourselves deploy these notions in understanding different phenomena, or seek to understand the claims of other writers.

Third, and building upon these earlier lessons, we need to overturn the received wisdom about "functionalism" in the presently dominant Standard Picture. We must keep firmly in mind that there is plausibly no single, metaphysically unified position in "functionalism." Nor there is a single set of concepts of "functional property," "causal role," and "realization" – regardless of what proponents of the Standard Picture blithely assume. Instead, as we have seen, we have a variety of metaphysical forms of "functionalism" using a diverse range of often proprietary notions of "functional property," "causal role," and "realization" – and such diversity again needs to be acknowledged and respected.

Heeding these lessons would finally allow us to leave the interpretive "fog" that has damaged many recent discussions of "functionalism(s)." Greater care about the variety of "functionalist" positions and concepts allows one to avoid flawed accounts of phenomena or misplaced critiques. We could thus finally discern which "functionalist" frameworks work best with certain phenomena and not others. For example, given the problems we have documented, one should avoid use of the Standard Picture of "functionalism," and its machinery of topic-neutral Ramseyfication and second-order properties, in understanding special science phenomena. Instead, we should favor the framework of Mechanistic Functionalism that, unsurprisingly given its genesis, provides a better treatment of the special sciences.

To conclude, following such an approach in different areas would transform the dizzying array of distinct versions of "functionalism," and their proprietary concepts, from a damaging trap for the unwary into a tremendous resource for the meta-methodologically aware philosopher. However, this is only the case if we leave the present "fog" that permeates many debates by keeping a clear appreciation of the diversity of "functionalism(s)" and their proprietary concepts of "functional property," "causal role," and "realization," as well as remembering the range of distinct research programs that use these notions to understand their very different object phenomena.[22]

References

Aizawa, K., and C. Gillett. 2009a. Levels, individual variation and massive multiple realization in neurobiology. In *Oxford handbook of philosophy and neuroscience*, ed. J. Bickle. Oxford: Oxford University Press.

Aizawa, K., and C. Gillett. 2009b. The (multiple) realization of psychological and other properties in the sciences. *Mind and Language* 24: 182–208.

Amundsen, R., and G. Lauder. 1998. Function without purpose. In Bekoff, Allen and Lauder 1998.

Bekoff, M., C. Allen, and G. Lauder (eds.). 1998. *Natures purposes*. Cambridge, MA: MIT Press.

Block, N. 1980a. Introduction: What is functionalism? In Block 1980b, 171–184.

Block, N. (ed.). 1980b. *Readings in philosophy of psychology*, vol. 1. Cambridge, MA: Harvard University Press.

Block, N. 1994. Functionalism (2). In Guttenplan 1994.

Buller, D. (ed.). 1999. *Function, selection and design*. Albany: SUNY Press.

Chalmers, D. 1996. *The conscious mind*. Oxford: Oxford University Press.

Craver, C. 2012. Functions and mechanisms: A perspectivalist view. In *Functions: Selection and mechanisms*, ed. P. Huneman. Dordrecht: Springer.

Cummins, R. 1975. Functional analysis. *The Journal of Philosophy* 72: 741–765.

[22] Thanks to Philippe Huneman, Ken Aizawa, Ron Endicott, Gualtiero Piccinini, and also to the audience at the 2009 SSPP conference in Savannah for comments on versions of the chapter.

Cummins, R. 1983. *The nature of psychological explanation*. Cambridge, MA: MIT Press.
Dennett, D. 1969. *Content and consciousness*. London: Routledge and Kegan Paul.
Dennett, D. 1978. *Brainstorms*. Montgomery: Bradford Books.
Fodor, J. 1968a. The appeal to tacit knowledge in psychological explanation. *The Journal of Philosophy* 65: 627–640.
Fodor, J. 1968b. *Psychological explanation*. New York: Random House.
Fodor, J. 1974. Special sciences: Or, the disunity of science as a working hypothesis. *Synthese* 28: 97–115.
Fodor, J. 1994. Jerry Fodor. In Guttenplan 1994.
Fodor, J. 2000. *The mind doesn't work that way*. Cambridge, MA: MIT Press.
Gillett, C. 2002. The dimensions of realization: A critique of the standard view. *Analysis* 62: 316–323.
Gillett, C. 2003. The metaphysics of realization, multiple realizability and the special sciences. *The Journal of Philosophy* 100: 591–603.
Gillett, C. 2007a. A mechanist manifesto for the philosophy of mind. *Journal of Philosophical Research* 32: 21–42.
Gillett, C. 2007b. Understanding the new reductionism: The metaphysics of science and compositional reduction. *The Journal of Philosophy* 104: 193–216.
Gillett, C. Unpublished. *Making sense of levels in the sciences: Composing powers, properties, parts and processes*.
Guttenplan, S. (ed.). 1994. *Blackwell guidebook to the philosophy of mind*. Oxford: Basil Blackwell.
Kim, J. 1992. Multiple realization and the metaphysics of reduction. *Philosophy and Phenomenological Research* 52: 1–26.
Kim, J. 1996. *Philosophy of mind*, 1st ed. Boulder: Westview Press.
Kim, J. 1998. *Mind in a physical world*. Cambridge, MA: MIT Press.
Levin, J. 2006. Functionalism. *Stanford Encyclopedia of Phil* (Accessed October 1st, 2006). http://plato.stanford.edu/archives/fall2006/entries/functionalism/
Lewis, D. 1966. An argument for the identity theory. *The Journal of Philosophy* 63: 17–25.
Lewis, D. 1972. Psychophysical and theoretical identifications. *Australasian Journal of Philosophy* 50: 249–258.
Lewis, D. 1994. David Lewis: reduction of mind. In Guttenplan 1994.
Lycan, W. 1987. *Consciousness*. Cambridge, MA: MIT Press.
Lycan, W. 1994. Functionalism (1). In Guttenplan 1994.
Pereboom, D. 1991. Why a scientific realist cannot be a functionalist. *Synthese* 88: 341–355.
Pereboom, D. 2002. Robust nonreductive materialism. *Journal of Philosophy* 99:499–531.
Piccinini, G. 2004. Functionalism, computationalism, and mental states. *Studies in History and Philosophy of Science* 35: 811–833.
Piccinini, G. 2010. The mind as neural software? Understanding functionalism, computationalism, and computational functionalism. *Philosophy and Phenomenological Research* 81: 269–311.
Polger, T. 2007. Realization and the metaphysics of mind. *Australasian Journal of Philosophy* 85: 233–259.
Putnam, H. 1960. *Minds and machines*. Repr. In Putnam 1975.
Putnam, H. 1973. *Philosophy and our mental life*. Repr. In Putnam 1975.
Putnam, H. 1975. *Mind, language and reality*. Cambridge: Cambridge University Press.
Rey, G. 1997. *Contemporary philosophy of mind*. Oxford: Basil Blackwell.
Shapiro, L. 2000. Multiple realizations. *Journal of Philosophy* 97: 635–654.
Shapiro, L. 2004. *The mind incarnate*. Cambridge, MA: MIT Press.
Shoemaker, S. 1980. Causality and properties. In *Time and cause*, ed. Van Inwagen. Dordrecht: Redial.
Shoemaker, S. 2001. Realization and mental causation. In *Physicalism and its discontents*, ed. C. Gillett and B. Loewer. Cambridge: Cambridge University Press.
Shoemaker, S. 2007. *Physical realization*. Oxford: Oxford University Press.
Van Gulick, R. 1983. Functionalism as a theory of mind. *Philosophy Research Archives* 4: 185–204.

Part IV
Psychology, Philosophy of Mind and Technology: Functions in a Man's World – Philosophy of Technology, Design and Functions

Artifacts and Organisms: A Case for a New Etiological Theory of Functions

Françoise Longy

Abstract Most philosophers adopt an etiological conception of functions, but not one that uniformly explains the functions attributed to material entities irrespective of whether they are natural or man-made. Here, I investigate the widespread idea that a combination of the two current etiological theories, SEL and INT, can offer a satisfactory account of the proper functions of both organisms and artifacts. (Roughly, SEL equates a function with a selected effect and INT with an intentional content). Making explicit what a realist theory of function supposes, I first show that SEL offers a realist theory of biological functions in which these are objective properties of a peculiar sort. I argue next that an artifact function demonstrates the same objective nature as a biological function when it is accounted for by SEL, but not when it is accounted for by INT. I explain why a dual theory of artifact functions admitting both INT and SEL functions is to be dismissed. I establish that neither INT nor SEL alone can account for all artifact functions. Drawing the conclusion that we need a new etiological theory of function, I show how one can overcome the apparent inevitability of INT for some artifact functions. Finally, I outline a new etiological theory of functions that applies equally to biological entities and to artifacts.

1 Introduction

Presently, there are two major etiological theories of function, the selectionist one (SEL) and the intentionalist one (INT).[1] Both theories deserve the label "etiological," which is associated with the theory Larry Wright propounded in the 1970s, because they take up Wright's thesis that attributing a function to an X may serve to explain its

[1] We take here a relatively abstract stance since there are different versions of SEL and INT.

F. Longy (✉)
Institut d'Histoire et de Philosophie des Sciences et des Techniques,
CNRS Université Paris I Sorbonne, Paris, France
e-mail: f.longy@orange.fr

P. Huneman (ed.), *Functions: selection and mechanisms*, Synthese Library,
Studies in Epistemology, Logic, Methodology, and Philosophy of Science 363,
DOI 10.1007/978-94-007-5304-4_10,

etiology. In other words, the function of X tells us why X exists now or why it is to be found in a particular location.[2] Etiological theories of function have become very popular in the last 30 years because they account for that which is specific to functions. Not only do etiological theories explain the role that functions may play in etiological explanations, they also offer a relatively straightforward account of the teleological meaning and the normative import that many functional attributions have. In fact, sentences of the form "X has function F" are often understood as meaning that X is there *in order to* do F. The etiological theory proposes a historical interpretation of the phrase "is there in order to" by focusing on the reason why Xs exist or are to be found in a determinate location.[3] It also offers an account of the normative distinction between properly functioning and malfunctioning items. A properly functioning item is one that can fulfill the function attributed to it. It has the physical capacity to do what its function demands, as is the case with a working windshield wiper or a healthy kidney. A malfunctioning item cannot fulfill the function attributed to it. This is the case with a kidney that cannot filter blood or a damaged windshield wiper that leaves most of the water on the windshield. By identifying functions with historical properties, etiological theories make it possible to separate the possession of the function from the possession of the corresponding capacity. Two items may both have function F, because they have the same relation to some historical fact (or series of historical facts), yet differ physically, so that one may possess the capacity to do F while the other one does not.

According to SEL, saying that X has function F amounts to saying that X is there because previous Xs have been selected for having done or produced F. A classical example in the literature is that of hearts having the function of circulating blood. It is because previous hearts have been selected for circulating blood that present hearts actually have the function of circulating blood (and not the one of producing a rhythmic sound). Although SEL was devised primarily to explain biological functions, it can also apply to artifact functions, as many authors have pointed out.[4] In the first case, it is natural selection that is at work; in the second case, it is some sort of cultural selection.

[2] SEL comes directly from the theory of function propounded by Larry Wright in 1973. This explains, in part, why what we call here "the selectionist theory" is often simply called "the etiological theory." There are actually several historical reasons to this identification of the etiological character with the selectionist one. First, the selectionist theories propounded by Millikan and Neander solved some serious difficulties the initial theory of Wright encountered. In particular, they offered the means to make a clear distinction between types and tokens, the absence of which resulted in a very problematic circular relation between cause and effect in Wright's original definition of function. Second, the focus has been mostly on biological functions where only SEL is needed. For a synthetic presentation of the history of the etiological theories of function, see the introduction of Buller in Buller (1999), 1–27 or Godfrey-Smith (1993).

[3] X is a multipurpose term here. It can be used to refer either to a type or to a token. Moreover, it can designate either an entity or a particular trait. The first ambiguity is common; it will be removed when necessary. The second one makes it possible to treat simultaneously the cases where a function is attributed to a type of entity (hearts which have the function to pump blood) and those where it is attributed to a feature possessed by a type of entity (being vividly colored which has the function of making peacocks' tails attractive to peahens). In the second case, "Xs" replaces a complex phrase such as "in peacocks, the feature of having a tail which is vividly colored."

[4] See, for instance, Millikan (1984, Chap. 1), Bigelow and Pargetter (1987, §III), and Griffiths (1993, §8).

INT concerns only artifacts. It is supposed to account for their proper functions. The proper function of an item is the one attached to it as a member of a particular artifact type. It is different from the use function an object may get from the occasional use it may be put to, such as when a pencil is used as a hairpin.[5] INT identifies the proper function of an artifact with a specific intention. Roughly, to say that artifact X has function F means, according to INT, that whoever created X or put X in some specific location did so thinking that X would do F.[6] An object will have, for example, the (proper) function of slicing potatoes if whoever designed this type of object did so for that purpose. In the last 30 years, the great majority of the literature on functions has concerned biological functions. When artifact functions were considered, it was usually *en passant.* If some form of SEL was not supposed to account for them, it was then taken for granted that the job would be done by INT.[7] Recently, functions of technological artifacts have been investigated for their own sake by what may be called the Dutch school in philosophy of technology.[8] The new and interesting insights that have resulted from this research don't change radically the situation, since the principle that artifact functions depend on intentions has not really been challenged. To be more specific, two members of this Dutch school, Vermaas and Houkes, have presented a new theory, ICE (intentionalist, causal role, evolutionist), which is much more elaborate than the basic and classical INT. However, ICE still gives the decisive role to the designer's intention.[9] As a consequence, my criticisms against INT will also be directed toward ICE as far as ICE is meant as a theory of function. Vermaas and Houkes have themselves changed their mind on that matter. From 2006 onward, they have stressed that ICE should be understood as an epistemological theory concerning rational function attributions rather than as an ontological theory about the nature of functions, while acknowledging however that the two questions cannot be totally disconnected.[10]

At present, whoever wants to account in an etiological spirit for all the proper functions of material entities is faced with three possibilities: to adopt SEL for all of them (possibility n°1) or to adopt a dual theory consisting of SEL plus INT in either one of two possible forms. Either SEL accounts for all the biological functions and INT for all the artifact ones (possibility n°2) or SEL accounts for all the biological functions and a part of the artifact functions, while INT accounts for the remaining part of artifact functions (possibility n°3).[11] I will argue that none of these three possibilities are acceptable. N°1 and n°2 have to be dismissed because neither SEL nor

[5] Biological functions are, in fact, proper functions. That is, they are functions attached to a type of organ or organic part, irrespective of the specific uses to which some particular items of the type may have been put by someone or other.

[6] For a survey and for a general but self-contained discussion of the major INT theories sustained in the second half of the twentieth century, cf. McLaughlin (2001), Chap. III.

[7] See, for example, Bigelow and Pargetter (1987), §III and Neander (1991), 462.

[8] See the web site of "The Dual Nature of Technological Artifacts" project (http://www.dualnature.tudelft.nl).

[9] See Vermaas and Houkes (2003).

[10] See Vermaas and Houkes (2006) and Chap. 11 (this volume, Sect. 2.3).

[11] Possibility n°4 (to account for all functions by INT) supposes to embrace a theological perspective as in modern ages.

INT alone can account for all artifact functions. This will be shown in the course of the argument against possibility n°3, which is the most challenging option and the one on which I will focus. Interestingly, some of the arguments against possibility n°2 will show that the usual separation between biological functions and artifact functions is quite arbitrary.

I will demonstrate the difficulties that a dual theory like n°3 encounters by focusing on INT. I will argue that INT implies a determinate antirealism about functions and that this antirealism is highly problematic because (1) it results in an untenable ontological duality since SEL is a realist theory; (2) it goes against the implicit conception of function that is revealed in our current use of the term.

To sum up, I intend to prove in this chapter the four following theses:

1. An etiological theory of function needs to be realist all the way through (for biological as well as for *all* artifact proper functions).
2. INT, despite its broad acceptance, is not a satisfactory theory for any sort of artifact proper function, no matter what version of INT we consider.
3. It is possible to understand the conditions an etiological theory has to fulfill in a way that does not make INT mandatory for any artifact proper function.
4. It is possible to devise an etiological theory that accounts homogeneously for all proper functions of material entities. (I sketch it at the very end of this chapter.)

2 Realism About Functions

What does a realist conception of functions consist in? To answer this, it is useful to reflect upon the range of existing positions. At one extremity, one finds the Hempelian thesis that functions are fictitious properties, like witches or medieval humors are fictitious entities.[12] According to Hempel, functional discourse is just a heuristic tool. So, all functional expressions must eventually be eliminated from our scientific theories (of course, this does not concern the mathematical homonym). In some usages, these expressions should be replaced by unproblematic ones, such as "necessary condition," and in others, they should just be eliminated without being replaced by anything specific.[13] At the other extreme, one finds SEL which considers functions as a special sort of property whose distinctive character consists in resulting from selective mechanisms. Biological functions, as SEL analyzes them, are objective properties of an historical type.

[12] To be fictitious and to be reducible are quite different things. "To be a lucky charm" is fictitious (in a good theory of the world, nothing will be equivalent to this property), but "to have a determinate weight" is by no means fictitious; it is just reducible by definition to mass and attraction. Hempel did not use expressions like "fictitious properties," but they help summing up his position.

[13] Cf. Hempel (1959).

In between these two extremes, one finds the systemic theory (SYS), which is the major challenger to etiological theories.[14] According to Cummins, who propounded SYS in 1975, functions concern parts in a system. The function of a part X is simply the capacity (or disposition) by virtue of which X contributes to the functioning of the system under consideration. X has function F thus means, X has the disposition to do F, and it is by doing F that X, a part of system S, contributes to what S is doing. So, according to Cummins, functions typically name physical dispositions.[15] Now, physical dispositions are commonly admitted to be legitimate properties. So, Cummins' functions are not, in this sense, fictitious properties. However, the use of the term "function" suggests that there are two sorts of physical dispositions: functional ones and nonfunctional ones. Yet, this difference is imaginary. It is merely one of perspective; it depends on what the theorist decides to consider as a system. So, in the end, functions constitute no particular subcategory of properties. Science can totally dispense with functions and replace them by the corresponding physical dispositions. This is why the etiologists usually see SYS as a *nonrealist* theory of functions. SYS doesn't recognize that something like fully fledged functions exist. It replaces functions by simpler properties that can ground neither an etiological assertion nor a normative one. On the contrary, philosophers more favorable to Cummins' approach judge SYS to be a realist theory since, unlike Hempel's, Cummins identifies functions with perfectly legitimate properties, physical dispositions.

At this point, I can make precise what I mean by a realist theory of functions. A realist theory of functions is one according to which functions are properties that are (1) not fictitious, (2) not replaceable by a simpler sort of property, and (3) objective. (1) and (2) have similar consequences: if (1) or (2) is not fulfilled, then functional sentences are specious; they deceptively support unwarranted etiological and normative inferences. So, a well-formulated scientific theory should admit functions only if (1) and (2) are fulfilled. (3) is also a condition required for admitting functions in science. Perhaps, with the exception of some branches of psychology, scientific discourse accepts only objective properties, that is, properties whose instantiation can be assessed by intersubjective means. Furthermore, as we will see in what follows, (3) also reflects the legitimacy conditions associated with the uses of "function" in everyday life.[16] Now, we are in position to determine precisely where SEL and INT stand relative to this realist/antirealist opposition.

---

[14] Cummins' theory is not considered here as a possible option because we share the judgment of many philosophers that it does not and cannot account satisfactorily for the normative and teleological aspects of functional attributions. We refer to it now just in order to make more precise what requirements a realist theory of function must meet.

[15] A physical disposition is a disposition which results from the physical properties of the item considered. The interest of a functional analysis is indeed for Cummins to offer a reduction to some lower level, the bottom levels being physical ones (see Cummins1973, §III.4).

[16] The expression "legitimacy conditions" indicates that the decisive point is not so much how we may in fact use a word as how we intend to use it. The fact that Bill had called a sheep a dog says less of what he thought "dog" might mean than the fact that he wanted to correct his previous statement when he saw better the animal and heard it bleat.

When applied to biological functions, SEL is realist. As we will see, it satisfies (1), (2), and (3). Up to the end of this section, our concern will be exclusively with SEL within the biological domain. It is clear that the biological functions SEL accounts for are objective and not fictitious. They are objective historical properties of features (or of organs). A function is a past effect of a feature that has helped those who bore the feature to survive longer or to reproduce more than those not bearing it. So, (1) and (3) are fulfilled by SEL functions straightforwardly. (2) raises the more difficult question of reducibility. Function F refers, according to SEL, to past facts (to past items that have done F), so it cannot be identified with a current physical disposition to do F as in SYS. But other reductions might be available. It could well be that functions understood as selected effects are reducible to a vast complex of past physical dispositions. If so, does the reducibility issue really mark a difference between SEL and SYS functions, since it cannot be shown *a priori* that functions are irreducible to physical dispositions?

Yes, it does. The decisive point is not whether something is reducible to something else, but whether the equivalents delivered by the reduction retain the salient aspects of the reduced elements and preserve their specificity. SYS denies reality to functions because the reduction it proposes implies the elimination of all that is supposed to be peculiar to functions. SEL, on the contrary, grants functions their peculiar features. In fact, the historical properties SEL proposes as equivalents of functions can support the etiological and normative sense of functional assertions. SEL, in itself, does not exclude the possibility of a further reduction. It leaves the question open, but it puts an important condition on reductions. Whoever adopts SEL should be willing to accept a further reduction to physical dispositions only if this reduction doesn't eliminate the peculiar features of functions. If ever such a reduction occurred, function F would be equivalent to a complex of physical properties and dispositions that would, no doubt, be very different from the simple physical disposition to do F. The specificity of functions would not disappear. In all likelihood, such complexes would be distinguishable from other sorts of complexes of physical dispositions and also from simple physical dispositions. In conclusion, SEL functions satisfy (2) because SEL excludes the possibility that functions could be replaced by simpler properties, that is, properties failing to support the etiological and normative assertions that functions support.[17]

Now, we can turn our attention to the functions of artifacts. In order to assess what kind of property the proper function of an artifact is, we need to make clear in what way artifact functions depend on human intentions and how significant this dependence is from an ontological point of view. The question to be addressed first is whether artifact functions should be counted among objective or subjective properties.

[17] See Huneman (Chap. 7, this volume) for a similar assessment of the difference between SYS (or causal-role functions) and SEL (or selected-effects functions). Huneman's "weak realism" about functions implies likewise that the difference between a mere effect and a function must consist in some objective fact in the world.

3 The Distinction Between Subjective and Objective Properties

Since the subjective-objective dichotomy can refer to many different distinctions, I need to specify the one that interests me here. Let me make clear, first, that my concern is with an ontological issue. Roughly, a property will be subjective if its realization depends *essentially* on some mental contents. It will be objective if its realization depends *essentially* on some objective facts. In order to give this abstract characterization a comprehensible content, it will be helpful to consider some examples. Suppose that commenting on some ice cream I am eating, I say:

1. I really enjoy its flavor.
2. It tastes like roasted nuts.
3. Most people did not like this flavor 10 years ago.
4. It contains at least 5% cream.

A perfect example of a subjective property is the property of having a flavor that the eater enjoys, which occurs in assertion (1). Property (1), as I shall call it, is a relational property. It bears on the relation between an ice cream and whoever eats it. In this case, it bears on the relation between ice cream B and me. Property (1) is subjective because its instantiation depends on what I feel while eating ice cream B. In other words, the truth maker of assertion (1) is the mental contents of the person who eats the ice cream. A perfect example of an objective property is property (4): containing at least 5% cream. When instantiated in B, property (4) is an intrinsic physical property of B. It concerns its physical composition. No subjective element such as a perception, a point of view, a judgment, or a feeling is in any way involved in making assertion (4) true or false. The truth maker of assertion (4) is an objective fact. We may turn now to (2) and (3), which are more complex cases.

Let us consider (2), first. Sentence (2) appears ambiguous. Depending on the context, it may be interpreted as meaning "I experience the flavor of roasted nuts while eating this ice cream" or as meaning "this type of ice cream shares some chemico-physical property with roasted nuts which makes it taste like roasted nuts to us humans." So, depending on the interpretation, "tasting like roasted nuts" will refer either to some impressions I have while eating ice cream B or to a natural property of a relational nature, that of sharing a determinate chemico-physical property with roasted nuts. In the first case, assertion (2) attributes to the ice cream a subjective property; in the second case, it attributes an objective one. It may seem strange, at first, that the same phrase may serve to refer to two properties so different from one another. There is a connection between these two properties, however, that explains this semantic ambiguity. Suppose that many ice creams of the same type as B (B-type ice creams) demonstrate the same subjective property as B when they are eaten; they taste like roasted nuts to the people who eat them.[18] In this case, there is

[18] B is used to name a singular ice cream as well as type of ice cream, those that have the same composition and form than B. Such ambiguities are commonplace, and they are most of the time unproblematic. For instance, "I really enjoy this taste" can be understood as being both about a singular ice cream and about a type.

an intersubjective agreement on how B-type ice creams taste. And this suggests a common cause, a natural property which explains why they taste similar to different eaters. This kind of shift from an intersubjective agreement to an ontological supposition feels very natural. It mostly goes without saying. As a matter of fact, it seems to me that the more we feel confident that there is – or would be, if we are confident that others would experience what we do – a broad agreement in considering that X tastes the same as Y, the more we tend to interpret "tastes like Y" as referring to a natural property of X. Besides, in most cases, the default interpretation of a sentence such as (2) seems to be the objective one. In fact, when we want a sentence such as (2) to be interpreted as a purely subjective judgment, we usually feel the necessity to stress it by adding "to me" or some similar expression. The case of sentence (2) is interesting because it demonstrates that determining whether a property is objective or subjective is not always obvious. Moreover, an expression may refer to an objective property indirectly by using some subjective property as an intermediary. That may happen, in particular, when it is supposed that there is some significant causal relation between the two.

There is a similar ambiguity in the case of (3). (3) could mean that most people who ate B-type ice cream 10 years ago did not enjoy its flavor or that most people who had tasted B-type ice cream 10 years ago did not subsequently buy the flavor very often. In the first case, the truth makers of (3) are felt impressions and, in the second case, observable behaviors. This ambiguity is similar to that of (2), even if not perfectly identical with it. However, (3) is noteworthy for another reason. It is a sociohistorical property. In this category, one finds properties which refer to a complex mix of (objective) public facts and (private) mental contents. Such is the case with the property of "being legally married." In order to have this property, one must have played a determinate part in a particular sort of public event – a wedding ceremony. And, for a set of public behaviors to instantiate a wedding ceremony, various public facts such as the existence of written laws or the presence of a civil servant playing a certain role must be realized. This in turn requires the realization of various private mental facts such as the facts that people *thought* of instituting such laws, that people *interpret(ed)* the laws, and, more important still, that the bride and groom *understand* what is said during the ceremony and what the institution itself means. In many countries, a marriage is invalid if the bride or the groom has been fooled into marriage and did not understand what was going on and what she/he was committing herself/himself to during the legal part of the ceremony. So, it looks as though the truth maker of a sentence such as "John is legally married to Janet" might be a mix of public (objective) and private (mental) facts. How should we then categorize the property of being legally married to Janet? According to our subjective/objective distinction, what is decisive is whether or not the realization of a property depends *essentially* on private facts and events. In order to understand what this means, one needs to know what the criterion is for subjectivity. Here it is. A property is subjective if there is a possible counterfactual world that differs from the actual world only with respect to mental contents and in which the property is not instantiated when it is instantiated in the actual world or vice versa.

The word "possible" reduces the domain of admissible counterfactual situations. The idea is simply that the change of mental contents you may envisage must be compatible with the world as we know it as far as public facts are concerned. Supposing that some individual who in fact enjoyed a determinate flavor did *not* enjoy it gives rise to a *possible* counterfactual situation. On the contrary, imagining that the thoughts of past legislators would have been totally different as far as marriage is concerned does not give rise to a possible counterfactual situation, since such a scenario is not compatible with the legal history that preceded John's wedding. Now, let us apply this criterion. It is possible to imagine that John, instead of thinking that he was really getting married, thought that he was acting in a comedy in which a wedding took place. Would that be sufficient to turn him into a bachelor? No, for a marriage to be annulled, much more is needed. To obtain that, John should, at least, make an official declaration, provide some proof for what he says, and enter a long and complex legal procedure. So, the property of being legally married to Janet is not after all subjective. Even if being legally married supposes the realization of a series of private events, it does not depend on private events in such a way as to be a subjective property.

Properties (2) and (3) are interesting because they have the same sort of complexity that some functional properties demonstrate. Now, they show that a property can be objective even though it is tightly connected to a subjective property (property 2) or even if it depends significantly on some mental events (property 3). A methodological conclusion can also be drawn from this discussion. In order to clarify what a property really refers to – and, in particular, whether it should be classified as subjective or as objective – the best way is to test upon what its instantiation depends.[19] And that can be done by considering counterfactual situations. Now, let us focus once again on functions.

There are object's functions which vary from one person to another. Consider a seascape painting that three people who share an office have agreed to hang. The artwork may have a different function for each of them: for Mary, it has the function of making her daydream; for John, the function of distracting his attention from a dirty spot on the wall nearby; and for Judith, the one of reminding her of last year's beach holiday. Each of the office mates had a different intention when agreeing to hang the picture on the wall, and each of them has different thoughts when looking at it. These three functions are subjective properties. In fact, by attributing other thoughts to each of the protagonist, one may obtain that the painting looses these

[19] A clear common vision of how the function issue should be addressed and resolved is still missing. For instance, the recent discussion between Thomasson (2007) and Elder (2007) on the nature of both artifact kinds and artifact functions shows an absence of common ground. On the one hand, Elder defends a realist position on artifact kinds and proper functions relying on an ontological analysis of copied kinds as natural kinds (pp. 35–40); on the other hand, Thomasson intends to show that "if function is what is relevant to membership in an artifactual (rather than natural) kind, it must be *intended* function that is relevant" since the creator's intentions "are most relevant to determining whether or not her product is in the extension of an artifactual term" (pp. 57–58 and see, in general, pp. 59–63). According to me, we can use this instantiation test to simplify the issue by settling some major points. In this way, one can establish which features of function really *need* to be accounted for.

three functions. However, these are not the sort of functions that are our concern here. They are use functions, and use functions need not be public. On the contrary, proper functions are attached publicly to an artifact or an organic part. They do not depend on what the artifact's owner thinks or does. For example, if Mary uses her Italian coffee machine as a decorative object and never uses it to make coffee, it nevertheless continues to be a coffee machine, that is, an object whose proper function is to make coffee. It is just that it has a use function on top of its proper function. Compared to proper functions, use functions are transient and superficial. So, despite obvious similarities, the nature of proper functions cannot be deduced from that of use functions.

According to INT, the proper function of a new trait (or entity) that results from human voluntary action is determined by intentional elements, specifically by the thoughts and intentions of those who are causally responsible for the coming to existence of the new trait (or entity). Let us consider various cases where human voluntary action plays a determinant role in the apparition of a new entity or a new feature, to see, first, when it appears appropriate to apply INT and, second, whether INT offers a good account of what the proper function is in some of these cases.

4 Human Intentions at the Root of a Biological Function

The first case I will consider is that of a biological trait which is the consequence of an intentional action. Imagine that some plant produces a new organic covering which is impermeable to DDT. In other words, imagine an evolution in which a plant, which reacted badly to DDT, gave rise to a new variety, which grows a covering preventing DDT from contacting some weak parts. Imagine also that this is a consequence of a situation that was produced voluntarily by some agents. They sprayed DDT for years in the new plants' range in order to avoid garden pests. What is the function of the new covering? The answer is obvious: to protect the plant from being damaged by DDT. The existence of this new covering resulted from human voluntary action and from evolution by natural selection. Since natural selection is involved, SEL should be able to account, totally or in part, for this function. According to SEL, as we have seen, a function refers to the relation that a trait has to a determinate series of historical facts, the ones which explain its current presence, that is, its selection or its maintenance. Some elements of the causal history leading to the current trait's presence will, accordingly, figure in the characterization of its function. Which ones? Those that are strictly needed to explain selection and maintenance: the favorable effect and the elements of the past environment that have made it selectively advantageous.

An example will help to explain and justify this assertion. Suppose that some animal species has a type of thick skin that can serve both for thermal insulation and for filtering a high level of UV radiation. The function of this thick skin will be thermal insulation, if the animals that possessed it survived and reproduced better than the ones that didn't because they suffered less from cold. Conversely, the function will be insulation from high UV radiation if what advantaged the first group of animals was that they suffered fewer cancers as a result of high levels of UV radiation. Depending on whether the function is thermal insulation or protection from a high level of UV radiation, the temperature or the amount of UV radiation in the past environment of the species will be involved in characterizing the function. The other elements of the causal story will not make a difference when it comes to determining the function of the thick skin. While some events, processes, or states of affairs may have played a decisive causal role in the development and maintenance of such a thick skin, the function of the thick skin does not depend as *directly* on them as it depends on the past temperatures or on the past levels of UV radiation. Indeed, any possible history that shares the pertinent part of the actual history would ground exactly the same function. Thick skin would be endowed with the function of thermal insulation through any history in which animals with thick skin survived and reproduced better than those with thin skin because the former were protected better against the cold.

Let's return now to our example of a new organic covering protecting the plant from DDT. Here, the intentions and the actions of the agents who have created the situation by introducing DDT in the soil are left out when one applies SEL. They belong only to the causal background. To account for the covering's function, one needs to refer only to the concrete situation produced by the intentional agents – the presence of DDT in the plant's environment – and nothing more because the favorable effect is just protection from DDT. This is confirmed by applying the criteria devised above. Imagining that the people who spread DDT did it in order to please some god (DDT being seen as an object of offering), or for whatever other reason, does not change a thing as far as the function of the protective covering is concerned. This is sufficient to prove that the thoughts and intentions of those who launched a biological evolution leading to a plant variety with an anti-DDT covering don't figure in the series of historical facts that constitutes the functional property. As this example shows, it does not necessarily matter whether or not intentions play or have played some decisive causal role in producing the functional feature. Functions may not depend at all on human intentions and actions even when functional features originate from human intentions and actions. In conclusion, in a case such as this one, INT has no role to play. The function is totally accounted for by SEL. Moreover, SEL's account helps understand why the originating human intentions and actions don't really matter.

One may wonder about the utility of taking as an example a function so different from a typical artifact function. First, it concerns a natural entity. Second, according to our story, nobody planned a transformation of the plant; it occurred as an unforeseen effect of spraying DDT in the area. Now, a typical proper artifact function is relative to a man-made device that has been planned for doing something specific, as is the case with a windshield wiper or a coffee machine. In fact, it came in the end as no surprise that the covering function was not intentional. Two remarks are appropriate here. First, the difference between our protecting-from-DDT function and a typical artifact function is not as big as it may seem. Voluntary and involuntary artificial selection of natural organisms may produce traits whose functions seem more similar to artifact functions than to biological ones. In industrial farm animals, traits hindering the capacities to survive and reproduce may be developed in order to fulfill some of our needs. Those traits have functions, such as making intensive farming easier. Such traits appear not biological, but artificial. Besides, as we will see below, there are artifact functions, or functions of man-made devices, that have not been planned by anyone. Second, to say it briefly, the study of non-paradigmatic cases helps to understand which features of the paradigmatic cases may deserve attention. What lesson can we draw from our protecting-from-DDT function example? In order to prove that a function includes some intentional component, it is not sufficient to demonstrate that it is the result of human actions and intentions.

5 Sociocultural Functions

Intentions play a much bigger role in functions that depend on some sort of sociocultural selection than in our previous function example. In the latter case, the selection pressure – to survive and reproduce in a context where there is DDT – could be described without any reference to intentions. Intentions cannot be left aside in a similar manner in the case of sociocultural selection. Let us consider cigarette holders, and for the sake of simplicity, let us take cigarette holders without any sort of filter. They have essentially a social function, that of making their bearers look sophisticated. Looking sophisticated is a social fact which depends mostly on intentional phenomena. Essentially, the selective pressure relative to this particular function calls into play the ideas people have formed about what it means to look sophisticated in what circumstances.

As explained earlier, a rough description of the situation is not sufficient to establish what sort of property a function is. In order to do that, we need to determine whether or not the instantiation of the function depends on some particular mental content. Let us address this issue by considering, first, those who are responsible for the coming into existence of the artifact. These are the individuals to which INT (the intentionalist theory of function) gives a leading role. Usually, the spotlight is on those who invented the artifact, but we may also include those who first produced the artifact and those who first distributed it commercially. According to the current intentionalist theories of function, the function of an artifact is determined by the mental contents of its creator(s). What he/she/they thought the artifact was made for is what its (proper) function is. So, our investigation will answer two questions at once about proper functions of the sociocultural sort – the type of function had by artifacts that are well known and widely diffused – first, whether they are subjective by depending on the mental content of some creators or introducers and, second, whether INT can account satisfactorily for them.

We can imagine the inventor(s), the first producer(s), or the first distributor(s) to have had whatever idea we want about the function of the cigarette holder. Their ideas about the cigarette holder's function are of no consequence as long as public facts remain the same. All other things being equal, their beliefs and intentions make no difference to the cigarette holder's function. If one of them or all of them thought that the distribution of the cigarette holder would mostly be limited to manual workers because they would have good use for it – better and cleaner handling of cigarettes with dirty hands or work gloves – and so would become a symbol of the working class, the function of the cigarette holders remains exactly what it is, a sign of sophistication.

The supposition that a significant change in the thoughts of those who invented an artifact could have no significant public effects may appear odd. However, it is not an implausible supposition. In order to show that, let us imagine two possible worlds. In the first one, which is either the actual world or one very close to it, a woman invented the cigarette holder thinking that it would turn smoking cigarettes into an elegant behavior. Then, she convinced someone to produce it and distribute it. In the second one, the counterfactual world, she invented the cigarette holder thinking it would be useful to manual laborers and convinced the same person to produce it and distribute it. In the first possible world, the cigarette holder comes into existence as an accessory for elegant men and women; in the second one, it is created as a handy device for manual laborers. However, the difference between the two worlds can stop there. Nothing prevents the two worlds from subsequently sharing the same history. In both worlds, the cigarette holder can begin its public life by being sold in bars where it is advertised by a stylized drawing of a man who smokes with it.[20]

We have focused on inventors, producers, or distributors because they are the only ones that might have the capacity to fix an object's function by thinking it to be designed for this or that. By enlarging the perspective and considering the history of the artifact, one can indeed see how thought may interact with a function's determination. But one also sees clearly that the relation between thought and function is indirect. If the function of an artifact depends on sociocultural history rather than on the mental contents of its creator(s), then what is decisive is what people do, not what they think. Let us be a little more specific.

If we consider all the people pertaining to the context in which cigarette holders are produced, distributed, sold, and used, then, of course, it matters what they think it is made for. If everyone thought of this very object as an acoustic device, it would not have the function it has. But then, this would show in many behaviors: instead of smoking with it, people would put it in their ears in determinate acoustic circumstances; they would decide to buy it when wanting or needing some sort of

[20] One can raise the objection that my example does not concern the "primary function", namely, to hold cigarettes, but some social secondary function. The same thought experiment can be carried out about primary functions, and it will deliver the same verdict. One could, for example, modify in that perspective a nice illustration of function change given by Beth Preston, that of the pipe cleaner that acquired the function of homecrafts or toys for children (Preston 1998, 241). One can imagine the object planned once as a pipe cleaner and once as a toy resulting in the same public situation: it came into use as a toy and not as a pipe cleaner. This could be possible by imagining in the two cases the same, not very clear, advertisement campaign where a grandfather is shown smoking the pipe and surrounded with children and where the device is represented as a funny being that twists itself in all manners.

acoustic device; etc. Treating something as an acoustic device may not be sufficient to make it an acoustic device, but it is surely sufficient to make it not be a cigarette holder.[21] So, the thoughts people entertain about an artifact indeed play a significant role, but the role these thoughts have is through the behaviors they cause – what one does and what one says. The behaviors contribute to the making of a general context, a sociocultural one. This context exerts a selective pressure on the production and distribution of artifacts such as cigarette holders. It is on this sociocultural context that the proper function of an artifact, its public typical use, *directly* depends not on ideas in minds. Besides, the fact that sociocultural contexts depend to a large extent on what people think does not make them any less objective than more natural contexts, such as the ecological niches of animal species. In fact, sociocultural facts and contexts are objects that sociologists and historians study with no problem using standard scientific methods.

So far, the application of our criterion has delivered nothing very surprising. As long as one can identify some social process similar to natural selection, one can apply SEL, and the functions so analyzed prove to be objective relational properties. The function of X, as analyzed by SEL, refers to *public* selected effects. Previous Xs have been used publicly in some determinate way and thereby have had definite objective effects. It is these past usages and these past effects that explain the existence of the current Xs and their being associated with a typical use. So, the application of SEL delivers the same realist picture whether it applies to biological functions or to artifact ones. With regard to functions, sociocultural selection is on a continuum with natural selection.

Not only does SEL apply well to many artifact functions, for many of them, it appears to be the best possible account. Even in the many cases where the publicly recognized function is exactly the one that was thought of by the inventors, it seems preferable to adopt a SEL account rather than an INT one (suppose for the time being that INT accounts are admissible). One reason for this is coherence in accounting for a change of function. As a matter of fact, one usually needs to bring in SEL in order to explain a change of function. To take Beth Preston's example, when something invented as a pipe cleaner and used for a while as a pipe cleaner gradually becomes used for homecrafts, only SEL can account for the second function because there is no inventor's intention or the like that could be referred to. Now, if the second function is accounted for by SEL, it seems SEL should also account for the first. If the sociocultural context is sufficient to fix the function in case n°2, it is also sufficient to fix it in case n°1. Moreover, the gradual change of function can be explained by a gradual change in the sociocultural context, but not by a discrete change of status. It cannot be explained by a jump from being an INT function to becoming a SEL function.[22]

What is more, such cases show that possibility n°2 (INT for all artifact function) *must* be rejected. Many artifact functions *cannot* indeed be accounted for by INT.

[21] We will show later on that it is, in fact, not sufficient.

[22] See also Longy (2009), §4, 61–62.

Among these are not only functions resulting from a gradual change in public use, as we have just seen, but also functions coming from a long history of unconscious improvements. For example, a determinate form of hammer's handle can have the function of making the hammer well equilibrated and easy to hold, but it may be that nobody never really calculated and planned this form. It may be that a series of small modifications from the first rough exemplars in the Paleolithic ages have been gradually selected without anyone ever planning consciously the ergonomic resulting form.[23] However, if SEL can offer a good account for many artifact functions, it cannot for all of them.

6 The Problem Raised by New Artifacts

There are at least two types of cases which cannot fit into the general scheme of natural or cultural selection:

(a) The first generation of a new type of artifact
(b) An artifact which is unique

A fundamental condition for applying SEL is missing: there are no previous items on which selection could have operated relative to the functional effect under examination.[24]

One could, of course, question the suppositions of novelty and uniqueness just made. There are no clear discontinuities enabling us to distinguish modifications or improvements within one type from the creation of a new type. What will determine whether a new spaceship made to reach Mars should be seen as a slightly modified exemplar of the previous spaceships that were made to reach the moon or as initiating a new type? There is no principle for answering such a question. The notion of a type (or to use a Millikanian notion, the idea of a reproductive family) is much more vague when artifacts are concerned than when organisms are.[25] However, even if there is no precise boundary between improvement and novelty, we do in fact distinguish the two and consider that some artifact functions qualify without ambiguity as new. So, the question remains of determining what it is that these supposedly new functions name. If the fuzziness of the distinction between sheer modifications and real novelties does not dissolve the "problem of new artifacts"; however, it casts doubt on the current supposition that one can divide easily functions

[23] Wright (this volume, 13–14) stresses this point. He also offers nice examples of unplanned artifact features which gained their function through sociohistorical selection. The more telling one is, probably, the presence of separate controls for front and rear brakes in motorcycles.

[24] See Vermaas and Houkes (2003), 265–266.

[25] See Lewens (2004) for a clear analysis of how the absence of a heredity mechanism affects the determination of types in the artifact case (Chap. VII, in particular, p. 141 sq.). And see Longy (2009), 64–65 for more developed examples.

resulting from sociocultural processes from functions resulting from planning and design. This point will be revisited below.

The attribution of a function to a new artifact has a peculiar feature that seems to force an INT analysis on us.[26] We usually make such attributions by relying on what those who have conceived the artifact say about it. Say Jane is a designer who has conceived a new device. When someone asks Jane what the device's function is, most likely he will accept her answer. If no sociocultural selection has taken place and if the function attributed to X is exactly the one given to it by its designer, then it seems that the function of X cannot but refer to what the designer thought X was for. However, this inference is too quick. One should not jump from the premise that no sociocultural selection has taken place to the conclusion that no objective selection of any sort took place and that SEL, in whatever form, cannot be applied.

Maybe there have been prototypes produced in the research period, and the function that Jane attributed to some part she designed, E, relies not on what she planned, but on empirical experience. Suppose nobody planned that this part should reduce vibrations, but in testing different prototypes, it was discovered that some prototypes fared better than others, and these were the ones that vibrated less. Suppose, then, that it was further found out that the prototypes that vibrated less were the ones with element E (instead of E′,E″...) in place P (we can assume that the plan left some degrees of freedom about what should be in P). In such a case, the basis for attributing to E the function of reducing vibrations is the selection of E (against E′,E″...) for its favorable effect through an experimental setting, not an explicit intention Jane had concerning E when designing the artifact. A current supposition until now has been that the separation between selective situations (where some objective mechanism of selection could be operating) and intentional situations (where only what the designer has in mind could be operating) was clear cut. The underlying assumption was that such a separation coincided with whether or not the artifact had a public existence and was thereby submitted to the selective pressures exerted by consumers. But the previous example shows that no such clear line of divide exists between the two sorts of situation. The use of prototypes in tests could be seen as an analogue of natural selection. As a matter of fact, it is not difficult to apply SEL in this case and to equate E's function with an objective property: the favorable effect for which prototypes having E have been selected over variants (with E′,E″, ... instead of E) in experimental context C.[27]

However, there are probably a lot of cases where no such selection of prototypes ever took place. Moreover, there are cases for which prototype selection is simply impossible. For instance, considering spaceships, one cannot test and observe in

[26] Here, we are on Vermaas and Houkes' home ground, that is, rational function attribution. Their investigation (Chap. 11, this volume) is somewhat parallel to ours. Both investigations challenge the common separations between biological and artifact functions and aim at a wide-ranging theory that covers both. However, Vermaas and Houkes explore the question from an epistemological point of view (rational attribution), while I explore it from an ontological point of view. Below, the connection between the two investigations is discussed and specified.

[27] See Longy (2009), §5, for a much more detailed analysis.

real conditions prototypes of devices planned to do something at a temperature exceeding the one that can be produced on earth. So, even if SEL can account for more functions than we realized until now, the conclusion remains that SEL is not sufficient to cover the whole domain of the proper functions of material entities. However, the fact that there might be objective SEL functions where no one was expecting them teaches us something. It shows how improbable a simple and clear separation between "SEL functions" and "INT functions" is. This becomes clearer still if one envisages not a simple device, such as a potato slicer, but some really complex artifact, such as a spaceship. The planning of a spaceship is an utterly complex process including many phases and loops, where many skills are put into use. At the end of this complex design process, whoever would try to determine the source of the functions attributed to the different parts and features would probably arrive at the following description: some have been invented at the table, others result from testing prototypes, and others still come from importation by incorporating already largely employed and tested artifacts, not to mention computers running evolutionary algorithms which add a further complication (is running an evolutionary algorithm a form of experimental testing, a form of intellectual reasoning, or something in between the two?).

So, if SEL and INT are the only options available, the domain of the proper functions of material entities appears quite gerrymandered if not completely confused. SEL and INT functions seem to mix more or less everywhere. One can find no clear border separating the domain or SEL functions from the one of INT functions. For this reason, INT and SEL functions should be ontologically similar. INT should identify functions with the same sort of relational property with which SEL identifies them.[28] In particular, since SEL equates functions with objective properties, INT should do the same. Does it? This is the question I investigate in the following section.

7 Can INT Functions Be Objective?

The definition of function as put forward in classical INT analysis is the following:

> The (first) function of X is what the designer, when realizing her project, has thought X to be meant for.[29]

This definition reduces functions to subjective properties. In fact, using our criterion, the function will change if we suppose that, all other things remaining equal, the designer's thought is different. Besides, as it has been acknowledged, such a definition has serious flaws. Following it, any crazy function some crazy inventor would endow her invented object with should be admitted as genuine. But, if somebody makes a strange object out of an old motorbike and declares that its

[28] See Longy (2007b) for a more substantial justification.

[29] Cf. McLaughlin (2001), 50–53.

function is to control the thoughts of the people nearby, we won't take her word for it. In fact, the situation described earlier, in which one believes as true the answer one obtains, needs to be filled in with restrictive conditions. This is the case only when we have good reasons to think that what we are told is rational and well grounded. Engineers of a well-respected firm, accredited members of institutional research laboratories, and user manuals edited by good companies, all these, are reliable sources we trust.

Thus, it appears that the definition of function should at least be modified so as to be restricted to cases of rational thinking.[30] With such a modification, one obtains indeed a good characterization of attributions of function to new artifacts, a characterization which conforms well to the attributions we are ready to make. However, these restrictions don't change the situation in relation to the objective/subjective distinction. So defined, functions would still be subjective properties, admittedly of a peculiar sort. They would be properties related only to rational thinking. This can be established once again by using our criterion: if more than one rational attribution of function is possible, which is often the case,[31] then the function of X could be changed simply by supposing a change in the designer's mind. Had she thought, on some rational basis, that element X would do G instead of thinking, as she did, that it would do F, the function of X would not be to do F, as it is, but to do G.

Yet, performing other thought experiments, we see that when we attribute a function to an artifact, we pretend to attribute to it an objective property, not a subjective one, whether or not our attribution is constrained by rationality. Suppose, for example, you believed like everyone else that X had function F, for example, to cool some part should it reach a definite temperature, because this was what some scientific theory implied. Suppose, further, that this theory is refuted and it appears that X could never have done that. Will we still say that X has function F because that is the function the designer of X gave it on a rational basis? Certainly not. This, I believe, is sufficient to show that a function cannot be equivalent to the rational functional attribution made by the designer at the end of his planning activity. It would certainly be possible to add some further condition to rule out this kind of

[30] Vermaas and Houkes (2003) require that the designer's attribution be justified, which amounts to the same. See Thomasson (2007, 62) for a more recent proposal along this line: in order to succeed in producing an artifact with the intended function, she requires that the creator has a substantively correct concept of the object she intends to make and that she succeeds at imposing on the object most of the features relevant to executing that concept.

[31] We have seen that simple devices like cigarette holders can be attributed different functions rationally. Of course, this may seem much more difficult with complex artifacts. What other function could be attributed to an Airbus A320 than being a plane? However, this does not speak against the general conclusion obtained considering more simple cases. It shows only that rational inferences might be very constrained when considering complex artifacts. But the possibility of more than one rational attribution exists also with highly technical artifacts. One can imagine, for example, a kind of plastic, specially planned to make lighter wings, that could also have been thought of for making water evacuation easier and quicker on wings in case of rain because of its low deformability.

counterexample, but I don't think this is a good strategy. The difficulty lies, it seems to me, in the very idea that a proper function could depend *directly* on mental content, that is, be a subjective property.

To be rational and well founded are epistemic properties, that is, properties relative to theories and knowledge, independently of what is the case. We can think that Ptolemaic astronomy was perfectly rational and well founded considering knowledge in ancient times and yet judge it to be absolutely false. Can proper functions be epistemic properties and depend on rational knowledge rather than on what is the case? Functions corresponding to epistemic properties would satisfy the following condition: if according to culture C, X has function F on a rational basis, then X has function F. But functions are not culture dependent in this way. Imagine, for example, a tribe whose theory is that solid things go under liquid ones which again go under airy ones. Suppose that to make a piece of wood or cork sink, these people attach to it a piece of lead that they have designed for this use. Suppose further that their explanation is that lead acts as a repellent for the airy parts which otherwise go inside the small holes that the wood and cork have. Let us add that they attribute explicitly to the pieces of lead they have manufactured the function to purify solids by repelling the air that otherwise would go in the small holes of the solid. Will we endorse their functional attribution, repelling air, since it is a rational one? No, we will not.

It may happen that an ethnologist or someone else who adopts our physics enunciates sentences that look like endorsements. Mary, a good ethnologist and physicist, might indeed say, "the pieces of lead have the function to repel air." The reason for this is that such sentences are ambiguous. They can be meant non-literally. You may not believe in lucky charms and still say "this is a lucky charm," meaning not that the object indicated brings luck or has the function of bringing luck, but that the people who had it thought that it did. For you, if this object has indeed a function, it is not to bring luck but to make superstitious people feel better. The way to test whether the enunciated sentence is meant literally or not is to imagine a dialog with a naive person, a child, for example. Suppose Mary's child asks her after having heard her speaking of the repelling function of the pieces of lead, "could I use one of them to repel the air that is in my mattress?" Mary's answer would probably run more or less like this: "Oh, they don't really have this function. Some people think they do, but, in fact, they don't. Their function is rather to make the things they are attached to become heavier, and that's how they make pieces of wood or cork sink in water" (one will not speak of Archimedes' principle or of specific gravity to a young child). We will not endorse the statement that some lead pieces have an air-repelling function unless we admit that at least, sometimes, some of them may have the capacity to repel air.

The conclusion I draw from these two thought experiments is that to have a certain function F, an artifact X must have *some objective property making it probable or at least possible* that it will have effect F in the right circumstances. It is not sufficient that there has been an attribution of function by the right kind of person (the designer or the producer), even if this attribution fulfills various conditions of rationality.[32] So, INT, classical or refined, is false. This conclusion, however, raises a new question: how is rational attribution of functions related to functions?

In other words, how can a theory such as Vermaas and Houkes' ICE deliver a satisfactory theory of rational function attribution but a quite unsatisfactory definition of function? The answer is simply that the attribution concerns epistemological conditions (what it is rational to suppose, admit, or judge according to our knowledge, our technical means, and the practical situation we are in), whereas the definition is concerned with ontological conditions (what the thing is like). Now, epistemological conditions and ontological ones can be quite different. The ontological condition for being gold is, we suppose, to have a certain atomic number, but the conditions for being rationally justified in thinking that something is gold are quite different. These can be seeing it labeled as gold in a jewelry shop; being told it is gold by a chemist; verifying it is yellow, malleable, and melting at k° in a determinate experimental setting; etc. The epistemological conditions depend heavily on the situation. They will not be the same for a customer buying jewels, a jeweler buying jewels, a trade expert on metals, or a chemist in his laboratory.

Of course, all epistemological conditions are not on equal footing. Those that correspond to expert knowledge deserve more consideration than others. Experts are called on when the best possible judgment is wanted. For gold, an expert is a chemist; for a new artifact function, an expert is an engineer involved in the designing process; etc. When there is an ontological hypothesis about what the thing is, the expert's judgment has to be coherent with this hypothesis, given the means at his disposal. Given our technical means and the ontological condition of having atomic number 79, it would be incoherent that appearing a determinate yellow be the epistemological criterion for the chemist. But coherence does not imply mirroring. Atomic number 79 can be the ontological condition even if the epistemological situation is rudimentary (no electronic microscope), and the best means at disposal to identify gold is a series of chemical tests such as the substance's reaction to heating or to being put in contact with mercury. Moreover, it is not even necessary to have a determinate ontological hypothesis to be able to distinguish between the two sorts of conditions.

So, even if the conditions of a rational attribution of function may be taken into consideration when investigating what functions are, they should however not be taken as a guide. The conditions for a rational attribution of function can be quite different from the ontological conditions fixing what a function is. This fact has two

[32] What is tricky with artifact functions is the fact that the designer's intention often seems to make all the difference between different possible functions. Two designers can conceive the same object for different purposes, and the two objects will normally not have the same function. One will have, let us suppose, the function of regulating the flow in some system and, the other, the function of eliminating some unwanted by-products in another system. But then, this difference will show objectively: the two objects will appear in different places or situations. Functions are relational properties implying contextual elements. So, it is to be expected that changing the context would change the function. It is the same with biological items: the tissue that has the function of filtering xyz in the kidneys might well serve a different function in the lungs. The difference between the natural case and the artifact one lies simply in the fact that the designer's by its action can determine in part the context.

important consequences. One is that an ontological theory of function may well leave room for different conditions of rational attributions of function depending on the domain. In particular, even if conditions for rationally attributing functions to organisms differ greatly from those for rationally attributing functions to newly made artifacts, the ontological condition for being a function may be the same in both cases.[33] Another consequence is that the situation can be much more intricate at the epistemological level than at the ontological one. So, the intricacy of the different sort of conditions for functional attributions does not prove that the notion of function refers to a jumble that should be eliminated or sorted out (a conclusion reminding of Hempel's). And, contra Cummins, it does not prove either that functional attributions designate something rather basic, such as physical dispositions, dressed in fancy functional clothes. This intricacy may result from varying epistemological conditions within a single domain. Various epistemological conditions attached to one ontological condition can indeed explain epistemological similarities that cross domains' boundaries. Indeed, there is nothing to be surprised at if engineers testing prototypes reason like evolutionary biologists or if biologists use reverse engineering techniques.

8 Is a Realist Etiological Theory of Artifact Functions Possible?

I draw one negative and one positive conclusion from the previous analysis: (a) the dual etiological theory must be rejected; (b) since no artifact function is subjective, one should account for all artifact functions with a *realist* etiological theory. More generally, I think we should aim at a unitary realist etiological theory that applies to biological functions as well as to all artifact ones. Some of the arguments presented here militate for such a unitary theory; there are still others that I have presented in other papers.[34] However, everyone wanting to go down that particular road faces immediately a serious difficulty. It seems at first sight impossible to give an etiological account of the functions of newly invented artifacts.

The etiological approach requires that the "expected effect" of Xs, that is, the function, intervenes in the causal chain that produces Xs. This sort of circular dependency between the existence of Xs and the capacity to do F can be achieved

[33] Of course, some of the arguments that militate for a unitary ontological theory militate also for a unitary epistemological theory. However, even in the case of two unitary wide-ranging theories, one ontological and the other epistemological, the conditions Q for rational attribution – it is rational to attribute function F to X when I know Q – will still greatly differ from the conditions P of possession; X has function F when P. At the epistemological level, identifying designing intentions is clearly decisive when considering artifacts, as Vermaas and Houkes (Chap. 11, this volume) correctly point out, at the ontological level; however, these intentions matter much less if they matter at all.

[34] See Longy (2007b, 2009).

by a selection mechanism, as SEL demonstrates well. However, there can be no such circular dependency in the case of newly invented artifacts. The capacity of Xs to do F cannot have had any causal action in producing the first generation of Xs, since no X existed before. The only possible story seems to be that the cause of the first generation of Xs was someone thinking that the Xs would have the capacity to do F. Thus, an intentionalist theory appears mandatory in this case. And, this explains why INT has been so widely accepted. However, we have seen that grounding functions on intentional content rather than on objective facts, as INT does, have the unwanted consequence of turning them into subjective properties. For a realist etiological theory of function to work for the first generation of artifacts, the cause of the existence of the Xs must be a relation between F and Xs that is in the external world and not only in the mental world of some individual.

Can we avoid having to choose between functions grounded either on an objective selection mechanism, or on an intentional content? It seems to me we can. We must just emphasize how much of invention is discovery. The idea that Xs will often enough do F in some set of circumstances may, when true (or true enough), be interpreted as a simple detection of an objective fact. Often, indeed, the thinking intermediary is made to disappear when objective facts are concerned. For instance, we simply say the cathedral of Strasbourg is 142 m high when we think that this common opinion is true. We make the intermediate thought manifest only when we discover errors. Then, we say something like "People once thought that the Strasbourg Cathedral was 142 m high." We favor objective relations over epistemic ones. In a somewhat similar manner, the thought that Xs will do F in circumstances C can be left out of the function picture. It can be seen simply as one of the steps through which the real relation between something's being of type *X* and being (sometimes or often) capable of doing F has given rise to the existence of Xs. No X need exist for someone to detect the objective relation between the Xs and the capacity to do F.

In fact, an etiological theory need not indicate some event in the causal chain leading to the Xs. What is required is an O such that "X is there because of O." Now, this does not imply that O should be a past event or a past state of affair. Nothing forbids O to be a timeless property – as is the timeless relation existing between the type *X* and the capacity to do F. So, the story may read somewhat like this: some intentional agent, let's say Andrea, who wanted something able to do F, decided to make Xs when she found out about O. In this case, O is indeed in an etiological relation to the Xs – the Xs are there because of O – it is just that O is not a material cause, but a reason. Such a description of the situation is moreover quite commonplace. For instance, let us suppose Andrea made a sharp-edged instrument thinking it would help her obtain the sort of holes she wants in felt pieces. Let us suppose also that what she thought was more or less correct; in most circumstances, her instrument in fact has this capacity. Then, we could indeed say that such an instrument was made because of its capacity to cut holes of a certain sort in felt pieces, without mentioning the role of Andrea's thoughts and intentions in producing the instrument. This solution to the problem of newly created artifacts needs to be substantiated, but it is contrary neither to good sense nor to our current thinking and speaking habits.

9 Conclusion: A Unitary Etiological Theory of Functions in Outline

Larry Wright thought of his etiological theory of functions as ranging over a vast domain, going way beyond that of biological functions. My project of a unitary theory for biological and artifact functions is reminiscent of this aspect of his theory, even if it is not as broad as his was. I don't want to account for the conscious functions that he considered at the end of his 1973 article.[35] The somewhat similar ambition explains another common feature between my tentative characterization, which I present below, and his own definition: the mechanisms producing the functions do not show up. Wright's definition of functions is more abstract than the ones propounded by defenders of SEL or INT. SEL's and INT's definitions eliminate the drawbacks of Wright's definition by restricting the domain of application and by introducing a specific mechanism, the one supposed to produce the typical functions of the domain under consideration. For biological items, this mechanism is natural selection. For artifacts, it is conscious planning. However, my account differs greatly from Wright's in many respects.

Let us consider first his definition of function. Over and above the technical flaws that doomed it from the start, it does not give a sufficient support to the so-called forward-looking aspect of functions.[36] It is the same for the SEL definitions that came after it, with one significant exception, the propensity version of SEL elaborated by Bigelow and Pargetter (1987). Of course, the etiological aspect of a classical SEL definition, which says that present Xs are there because previous Xs have done F, gives a good basis for inferences of the following sort: if some or most previous Xs were able to do F in some sort of circumstances or other, then some or most present Xs should be able to do so, too. Yet, it is one thing to offer the basis for inferring A, when the right sort of information is available (many traits are hereditary through genetic transmission, past and present environments are similar in pertinent features, and so on), but quite another thing to consider A as part of the definition of function. Function attributions are used heavily to give information about what is to be expected from some object or some feature. We name artifacts mostly after their function – potato peeler, hair dryer, coffee machine, etc. Our practical interests justify this naming practice; our concern is mostly with what the things are supposed to do, not with precisely what their makeup is. It is for their supposed capacities that we usually buy and keep artifacts. With organs too, the accent is on function rather than on physical makeup. Not much attention is paid to physical differences among organs when those differences do not affect the fulfillment of their function(s). For this reason, a

[35] Wright (this volume) does not renounce his ambition of an encompassing theory of function's attribution, but he acknowledges the necessity to emendate his former doctrine in order to account for the differences between function-patterns that are attached to kinds and function-patterns that concern the actions of individual agents.

[36] For a little more precision on the technical flaws, see above note 2.

characterization of functions that would include both an etiological aspect – why there are such things as Xs – and a forward-looking one, the capacities to do F of the present Xs, would be more satisfactory than a purely etiological one.

Even if Wright did not put any particular mechanism in his definition of function, he contended that any function of whatever sort resulted from selection in one form or another. If it did not result from an objective selection in the world, then it resulted from a mental selection in the mind.[37] The idea that conscious selection could be taken as a source of functions has been aptly criticized by Mitchell.[38] The main reason to reject it is simply that such "conscious functions" would be totally subjective. The function attributed by the crazy inventor would then be a function. If functions are objective properties, as I claim they are, selection cannot be the source of all of them. Yet, there is another way to maintain Wright's idea that for every function, the consequences must have played a decisive role in producing the functional items.[39] It can be done, as I have suggested, by isolating the part of discovery in invention. In this way, the consequences or the effects can play roles as *reasons* even if they don't play roles as material causes.[40]

I cannot justify here the new characterization of function that I have begun to explore and that I will give now in order to show that the critical analysis presented here does not lead to a dead end, but opens a new line of investigation. Let us quickly consider some of its salient features. Its most surprising one will certainly be, for a classical etiologist at least, the introduction of probability. To be specific, it introduces the probability that an item has to function properly, that is, the probability it has of having the capacity to do F in the right circumstances. Thanks to this incorporation of probability, the forward-looking aspect of functions is taken into account.[41] Moreover, the fact of linking every function to a specific probability rooted in a set of causes that make up a sort of essence (as I explain below) helps make the characterization substantial, since it is a compelling condition. It also helps understand why functions are so pervasive in science; they can serve to explain many statistical regularities that are nonarbitrary. For now, I have, for the most part, explored what the probability to be well-functioning refers to. My objective has

[37] See Wright 73, 50 (in Buller).

[38] See Mitchell (1989), 218 sq.

[39] Wright (1973, 162) writes, for example, "Why is it there?" in some contexts and "What does it do?" in most, unpack into "What consequences does it have that account for its being there?"

[40] As it shows still more clearly in the characterization below (see, in particular, condition 3), the etiological condition so understood can apply to historical functions (typical selected effects function) as well as to nonhistorical ones. So, acknowledging, as Bouchard (Chap. 5, this volume) does, that there are both historical and nonhistorical functions does not compel us to be a pluralist about functions, contrary to what he contends.

[41] I side with Bigelow and Pargetter (1987), Wimsatt (Chap. 2, this volume, 26), and Bouchard (Chap. 5, this volume, 62) on this issue. Against the classical etiological approach, which is totally backward looking, I propose, as they do, to introduce a forward-looking element in the guise of a probability (or a propensity) in the definition of functions. It should be noted however that rather than a fitness measure or something of the sort, I introduce the probability of having the capacity to produce the functional effect.

been to show that connecting each function with a determinate probability of having the corresponding capacity has a solid ground and is in no way a technical trick. I have argued that a definite probability to do F is, in fact, rooted in what makes the functional type considered not an arbitrarily defined class, but a *real kind*.[42]

The introduction of the notion of "real kind" is the other novelty that our characterization brings in. Biological species are real kinds, but real kinds include many more classes. Every class whose unity, going beyond an explicit listing of the common features of its members (its nominal definition), is rooted in physical, historical, or relational properties is a real kind. Thus, the category of real kinds includes physical natural kinds, as are, for example, gold or water, but it includes also the historical and homeostatic kinds of Ruth Millikan and Richard Boyd. Biological species as well as many current artifact types are such historical real kinds. They are kinds whose members share a common destiny because they depend equally on a series of common historical causes and mechanisms (heredity by genetic transmission, natural selection, cultural selection, common design, common manufacturing procedures, and so on).[43]

As a last remark, let me explain why I aim at a characterization, which is more modest than a fully fledged definition of function. The difference between a characterization and a definition is the following: a definition gives a necessary and sufficient condition, while a characterization delivers only a necessary one. To obtain only a necessary condition is unsatisfactory when the condition is evident and not very substantial. It delivers then little information if any at all. However, when the necessary condition is substantial and opposes present alternative definitions, the characterization is quite informative. I don't aim at a proper definition because I suspect that the sufficient condition will involve pragmatic elements without great significance. Whenever a regular connection between the members of some type *X* and some effect *F* can be explained by applying classical laws of nature, it will usually be explained this way. Attribution of functions and functional explanation are useful only when they are unavoidable, that is, when the regularity cannot be explained simply by resorting to the natural laws of physics or chemistry. As a matter of fact, one objection to classical SEL definitions has been that they apply to unwanted cases, for example, to some effect of crystals, even though we have never called such an effect a function and are quite reluctant to do so.[44]

Such objections are serious only for a conceptualist project whose aim is to clarify what we mean. The conceptualist has to give an explicit definition of the notion under scrutiny, a definition that reflects its actual conditions of use. As far as I am concerned, however, the main task of the theory of function is not to grasp the notion as a purely mental or conceptual entity, but as a notion that refers to some real feature in the world. Definitions and characterizations seen in this perspective are not meant simply to clarify our thoughts or ways of talking but also to elucidate what the thing

[42] See Longy (2006, 2007a).

[43] See Boyd (1989, 1991) and Millikan (1999).

[44] See Bedau (1991).

referred to is like. I need not argue here for this understanding of much of the conceptual labor of philosophers and scientists (certainly a good example of it is when biologists discuss what a species is); others have already done it.[45] Given my preference for an ontological clarification (what a function is) over a psycho-semantic clarification (what our present representations and conceptions of functions are), a characterization is good enough as long as it tells us what the specific features of functions are that make referring to functions a necessary means for describing and explaining a significant part of reality.

As a conclusion, let me give the tentative characterization of a proper function that I am aiming to explore further:

Item *I* has proper function *F* – the proper function to do F in circumstances C – only if *I* is a member of a *real kind* X such that:

1. There is a probability p ($0<p<1$) such that *X*'s members have probability p of possessing the capacity to do *F* (in circumstances C).[46]
2. The probability p of doing F that *X*'s members have is determined by the particular set of causes and conditions that makes *X*'s members form a determinate real kind.
3. *I* is there because of conditions 1 and 2, that is, because of the probability to have the capacity of doing F that (real and potential) *X*'s members have.

References

Bedau, M. 1991. Can biological teleology be naturalized? *The Journal of Philosophy* 88: 647–655.

Bigelow, J., and R. Pargetter. 1987. Functions. *The Journal of Philosophy* 86: 181–196.

Boyd, R. 1989. What realism implies and what it does not. *Dialectica* 43: 5–29.

Boyd, R. 1991. Realism, anti-foundationalism and the enthusiasm for natural kinds. *Philosophical Studies* 61: 127–148.

Buller, D. 1999. *Function, selection, and design*. Albany: SUNY Press.

Cummins, R. 1975. Functional analysis. *The Journal of Philosophy* 72: 741–765.

Elder, C. 2007. The place of artifacts in ontology. In *Creations of the mind: Essays on artifacts and their representation*, ed. S. Laurence and E. Margolis, 33–51. Oxford: Oxford University Press.

Godfrey-Smith, P. 1993. Functions: Consensus without unity. *Pacific Philosophical Quarterly* 74: 196–208.

Griffiths, P.E. 1993. Functional Analysis and Proper Function. *The British Journal for the Philosophy of Science* 44: 409–422.

Hempel, C.G. 1959. The logic of functional analysis. In *Symposium on sociological theory*, ed. L.Gross, 271–307. New York: Harper and Row.

Lewens, T. 2004. *Organisms and artifacts*. Cambridge: MIT Press.

Longy, F. 2006. Function and probability: The making of artefacts. *Techné: Research in Philosophy and Technology* 10(1): 71–86.

[45] See among many others, Boyd 91, 141.

[46] It can be that what is needed is a slightly more complex condition, that of having a probability within a determinate range $[q, q+\varepsilon]$, such that $0 \leq q \leq p \leq q+\varepsilon \leq 1$, but we need not consider such subtleties here.

Longy, F. 2007a. Two probabilities of dysfunction and two kinds of chance. In *Probability and causality in the sciences*, ed. J. Williamson and F. Russo, 335–360. London: College Publications.

Longy, F. 2007b. Unité des Fonctions et Décomposition Fonctionnelle. In *Le tout et les parties dans les systèmes naturels*, ed. Thierry Martin, 89–97. Paris: Vuibert.

Longy, F. 2009. How biological, cultural and intended functions combine. In *Comparative philosophy of technical artefacts and biological organisms*, ed. P. Kroes and U. Krohs, 51–67. Cambridge, MA: MIT Press.

McLaughlin, P. 2001. *What functions explain*. Cambridge: Cambridge University Press.

Millikan, R. 1984. *Language, thought, and other biological categories*. Cambridge: MIT Press.

Millikan, R. 1999. Historical kinds and the 'special sciences'. *Philosophical Studies* 95: 45–65.

Mitchell, S. 1989. The causal background for functional explanations. *International Studies in the Philosophy of Science* 3: 213–230.

Neander, K. 1991. The teleological notion of "Function". *Australasian Journal of Philosophy* 69: 454–468.

Preston, B. 1998. Why is a wing like a spoon ? A pluralist theory of function. *The Journal of Philosophy* 95: 215–254.

Thomasson, A. 2007. Artifacts and human concepts. In *Creations of the mind: Essays on artifacts and their representation*, ed. S. Laurence and E. Margolis, 52–73. Oxford: Oxford University Press.

Vermaas, P.E., and W. Houkes. 2003. Ascribing functions to technical artifacts: A challenge to etiological accounts of function. *The British Journal for the Philosophy of Science* 54: 261–289.

Vermaas, P.E., and W. Houkes. 2006. Technical functions: A drawbridge between the intentional and structural nature of technical artifacts. *Studies in History and Philosophy of Science* 37: 5–18.

Wright, L. 1973. Functions. *Philosophical Review* 82: 139–168.

Functions as Epistemic Highlighters: An Engineering Account of Technical, Biological and Other Functions

Pieter E. Vermaas and Wybo Houkes

Abstract In this contribution, we explore whether the ICE-theory of technical functions can be used to formulate a unified account of functional discourse in biology and other functional domains. We discern three routes for arriving at a unified account: literally applying the ICE-theory to the other functional domains, taking non-technical functions as 'as-if' ICE-technical-functions, and generalising the ICE-theory to the other domains. We argue that the first and second routes are rather unattractive; the ICE-theory presupposes descriptions of using and designing that cannot be literally applied to biology without counterintuitive results. The third route towards unification leads to a unified ICE-like function theory, but one that calls for reservation. The unified ICE-like function presents a general understanding of functional descriptions as descriptions of items by which agents are epistemically highlighting the capacities that explain (successful) realisations of goal-directed patterns designated by (other) agents. Yet this understanding contradicts the usual view that biological functions are features that biological items have independently of any goal-directed patterns designated by agents.

1 Introduction

One feature of functional discourse that makes it strongly appealing for philosophers is its sheer scope. Function ascriptions to objects and other functional descriptions are found in disciplines as diverse as biology, engineering, cognitive science and sociology,

P.E. Vermaas (✉)
Department of Philosophy, Delft University of Technology, Delft, The Netherlands
e-mail: p.e.vermaas@tudelft.nl

W. Houkes
Section of Philosophy and Ethics,
Eindhoven University of Technology, Eindhoven, The Netherlands
e-mail: w.n.houkes@tue.nl

P. Huneman (ed.), *Functions: selection and mechanisms*, Synthese Library, Studies in Epistemology, Logic, Methodology, and Philosophy of Science 363,
DOI 10.1007/978-94-007-5304-4_11,

to give just four examples. Consequently, many philosophers who have analysed the notion of function, perhaps focussing on one discipline, have felt the urge to export their approaches and results to other disciplines or domains. Thus, a unified account of all functional discourse acts as a regulative idea of philosophical analysis.

This is not to say, however, that all authors underwrite this idea – only that virtually all philosophers who analyse functional discourse have commented on it and have allowed the idea to guide their work. There are, in practice, at least three ways in which the regulative idea is studied or implemented. Firstly, authors may have arrived at an analysis of, say, biological functions and then go on to generalise it in a generous sweep to a theory that is supposed to hold literally for all types of functions and in all domains in which we find functional discourse. Others may be of the opinion that functional discourse, although it is used in different domains, is only truly at home in the domain of technical artefacts, i.e. those material objects that are used, and perhaps also designed, for practical purposes. This second route may still lead to a general, unified analysis, but one in which functional discourse outside of the domain of artefacts is 'as-if': by describing objects in these other domains in functional terms, they are considered *as if* they were technical artefacts. And thirdly and finally, there are authors who export approaches and results from one domain to another for defending that a unified theory is feasible in this more liberal sense or for showing that unification is as yet or forever an unattainable ideal.

In this contribution, we subject our own analysis of technical functions to the regulative idea of unification across 'functional domains', i.e. we explore which of the three routes discussed above are open for the ICE-function theory as presented in Houkes and Vermaas (2004, 2010) and Vermaas and Houkes (2006). Implicitly, the unification idea has been present in our earlier work on this theory. For this work was prompted by a critical third mode of export: we argued that the etiological approach towards biological functions cannot yield an adequate analysis of technical functions, thus raising the question of which analysis *is* adequate for the domain of engineering (Vermaas and Houkes 2003). The ICE-analysis of technical functions is our answer to this question. Yet, we presented this analysis for the engineering domain only, thus raising the further question of whether the ICE-function theory in turn can be taken as an adequate analysis of functional descriptions in biology and other functional domains as well. In this contribution we take up this second question. We argue that our theory makes the first and second ways of implementing the unification idea rather unattractive: the ICE-theory presupposes descriptions of using and designing that cannot be literally applied to, say, biology without considerably counterintuitive results – even when they are taken as 'as-if' descriptions. That leaves our exploration with the third route towards unification. The results this route leads us to do not warrant, however, a claim to unconditional success. We arrive at an ICE-like theory for biological functional descriptions and at a unified ICE-like function theory applicable to all functional domains; we can thus demonstrate that the ICE-theory for technical functions can be exported across functional domains, without claiming that the theories we formulate are the only ones obtained by exporting the original ICE-theory of technical functions. An assessment of the results calls, however, for reservation. The unified ICE-like function theory we arrive at has the *prima facie* advantage that it presents a general understanding of functional descriptions: it casts them as descriptions of items by which agents are epistemically highlighting the capacities that explain (successful) realisations of goal-directed patterns designated by (other) agents. On closer inspection, this understanding seems, however, a disadvantage in biology: biological functions are usually taken as features that biological items have independently of any goal-directed patterns designated by agents. Hence, our overall conclusion is that, if one accepts the ICE-theory for the technical domain, a unified function theory only seems possible if one accepts this disadvantage of the ICE-like theory for biological functional descriptions. If one is willing to pay this price, the third route towards unification is open to the ICE-function theory; if this price is deemed too high, the conclusion is that also our theory does not bring us closer to the ideal of a unified function theory.

We start, in Sect. 2, by briefly presenting the key elements of the ICE-theory for technical functions. In Sect. 3, we consider how this theory can be applied to biological functional descriptions. We argue against applying the ICE-theory literally to biology and against taking biological functions as the technical functions that biological items would have if they were taken as technical artefacts. Then, we explore how the ICE-theory can be transformed into a 'bio-ICE-theory'. In Sect. 4, we export the ICE-theory to a unified theory for all domains and state our conclusions.

2 The ICE-Theory for Technical Artefacts

The ICE-theory provides an analysis of functional descriptions in technology. It presupposes an action-theoretical description of the use and design of technical artefacts and characterises functional descriptions as identifications by agents of those physicochemical capacities of artefacts that make the use of these artefacts successful. We start by introducing the action-theoretical background. Then, we present the ICE-theory itself, and we argue why this theory is adequate for technology.[1]

2.1 *Using and Designing*

The action-theoretical background of the ICE-theory consists of a description of using and designing technical artefacts, and of the relation between these activities, in terms of use plans. This description is relatively straightforward: artefact use is the execution of a use plan for the artefact; designing is the development of those

[1] Sections 2.1 and 2.2 are drawn in part from Sects. 1 and 2 of Vermaas and Houkes (2006). See Houkes and Vermaas (2010) for a more detailed introduction and motivation of the ICE-theory of technical artefacts.

use plans for artefacts. The notion of use plan builds upon more general action-theoretical work (Bratman 1987; Pollock 1995), in which all intentional actions are described in terms of plans, i.e. goal-directed series of considered actions. We define a use plan *p* for an artefact *x* as a goal-directed series of considered actions in which manipulations of *x* are included as contributions to realising the given goal (Houkes et al. 2002; Houkes and Vermaas 2004). This is a rough-and-ready definition, but it is sufficiently precise for characterising functional descriptions of artefacts.

Consider, as a simple example, using a bicycle. Suppose someone wants to visit a friend who is staying elsewhere in town. Realising this goal involves planning, i.e. deliberating about a sequence of actions to be taken in order to realise the goal. This deliberation typically involves a choice of means. Let us say that, in this case, one could either walk or cycle to the friend. Both of these choices lead to different plans: different actions to be taken, a different route, etc. The first means, walking, consists of a number of actions (turning corners, looking left and right when crossing a street), which largely involve only the manipulation of the walker's own body. The second choice of means, cycling, leads to a plan that minimally includes manipulations of the bicycle. We call this series of considered actions – roughly, fetching the bike, stepping on it and pushing off by starting to rotate the pedals with one's feet – a use plan for the bike, and the use of the bike the carrying out of this use plan. Use plans do not come on a one-per-artefact basis, except perhaps for very complicated artefacts such as car factory assembly lines. Simple objects such as a piece of wire can be manipulated in different ways and for different goals, all giving rise to different use plans for the wire. A bicycle can be used for exercising as well as for transportation. In this case, the actions taken are largely similar, but the goals are different, giving rise to different use plans for the bicycle.

In deliberating about how to realise a goal by manipulating an object, an agent settles on a use plan for that object, but she/he need not execute it immediately, or even execute it herself/himself. Human beings routinely provide each other with plans through communication and instruction, verbal and non-verbal. Few of us are self-taught cyclists, and we are regularly instructed about how to realise familiar or new goals with newly introduced artefacts. In this way, designing can be included in our action-theoretical background.

If we take using as executing a use plan, we can characterise designing as developing a use plan and possibly describing the types of artefacts that are to be manipulated in the course of executing the plan. This turns designing into a primarily use-oriented activity: the goal of designing is to aid prospective users in realising their goals, by developing appropriate sequences of actions to be undertaken by the users. This user-aid goal makes the communication of use plans into an integral part of designing: this activity would be pointless if designers did not make their newly developed use plans available to users. In practice, there are various means of communication available, such as manuals, personal instruction, television ads and product demonstrations.

Our characterisation of designing is partly revisionist in the following sense. On it, designing includes some activities that are not regularly regarded as such. If an agent develops a creative way of using an existing artefact, and communicates this alternative use plan to others, she/he is – on our characterisation – involved in designing. Creating artefacts or blueprints thereof, usually regarded as the paradigm of designing, is for us merely a special case, henceforth called 'product designing'. Frequently, but not necessarily, developed and communicated use plans involve the manipulation of artefacts that are not available. In such cases, designers ought to contribute to the availability of these artefacts by describing and, possibly, creating them. Those subsidiary activities we take as product designing: designing is the development and communication of new use plans, and product designing is the description of the artefacts that have to be created for enabling the executing of the developed use plans. This difference between designing in general and product designing is one tool for characterising the roles of agents who, intuitively, are both using and designing artefacts; they design but do not engage in product design.

A third role, besides those of using and designing, is available for agents: that of observing artefacts. This non-practical role occurs less regularly in practice but is relevant to our function theory. In our analysis, when ascribing functions to artefacts, an observer constructs a use plan that she/he has neither designed nor learnt about via communication from other agents. This construction can proceed in two ways: an observer either believes that the use plan has been developed and communicated by its original designers – archaeologists typically hold such beliefs – or she/he believes that the plan has *not* been developed and communicated by its original designers – as analysts of artefact failures typically do.

2.2 *The ICE-Function Theory*

The ICE-theory primarily concerns functional descriptions of artefacts. In the theory, such descriptions are identifications by agents of the physicochemical capacities of the artefacts that make the use of these artefacts successful. With the action-theoretical analysis in hand, we can construct a more precise characterisation: agents describe artefacts functionally relative to use plans for these artefacts by identifying the physicochemical capacities of the artefact that make this use plan a successful means for realising the plan's goal. One immediate consequence of this characterisation is that it does not make sense to speak about technical functions of objects that are not, metaphorically speaking, embedded in a use plan. Agents identify capacities as the functions of artefacts only relative to use plans for those artefacts and, moreover, identify these capacities relative to evidence that justifies the identification. We refer to this evidence as an 'account *A*', following Robert Cummins (1975). In practice, this account may consist of experience of agents with artefacts and their use plans, of testimony provided by designers, of scientific and technological knowledge or of a combination of these sources of evidence.

Further specifications of the key elements of our characterisation lead to the following central definition of functional descriptions of technical artefacts:

An agent a justifiably ascribes the physicochemical capacity to ϕ as a function to an artefact x, relative to a use plan p for x and relative to an account A, iff:

I. a believes that x has the capacity to ϕ.
   a believes that p leads to its goals due to, in part, x's capacity to ϕ.
C. a can justify these two beliefs on the basis of A.
E. a communicated p and testified these beliefs to other agents, or a received p and testimony that the designer d who developed p has these beliefs.

This definition is normative. It covers many examples of actual function ascriptions but certainly not all.[2] We evaluate the latter cases as groundless function ascriptions.

Let us quickly review the norms on function ascriptions introduced in this central definition. We require functions to be ascribed relative to use plans.[3] We require, in the I- and E-conditions, a function-ascribing agent to have two particular beliefs, which may be based on three different sources of evidence. However, the norms inherent to the E-condition may need some spelling out. It requires, for starters, that designers d of use plans communicate to prospective users that they have produced or selected certain artefacts for the capacities corresponding to their functions. In addition, this communication must be successful, i.e. it must provide the prospective users with testimonial evidence for the beliefs that the artefacts have been produced or selected for the capacities corresponding to their functions. So if the agent a in the central ICE-definition is an agent who received this communication from this designer d, agent a can use this communication as testimony to justify his/her beliefs as required by condition C. Condition E captures, moreover, the intuition that there is, to some extent, a privileged (designer) perspective from which functions are ascribed: a function is a capacity that is selected by someone, presumably for good reason, and that is communicated to others, presumably to aid them in dealing with the thing in question. This does not mean that we require function ascriptions to refer to the activities of engineers or other technological professionals. On our characterisation of designing, any agent who develops and communicates a use plan and who can justify it, if only by plain experience that it works, is a designer and shares the privilege: we do not introduce social standards for preferring one kind of designer to another. So the agent a in the central ICE-definition may also be a creative user of the artefact x, who on the basis of his/her findings justifiably ascribes a function to the artefact and provides others with testimony for doing so as well.

[2] Function ascriptions by physicians to placebos do not satisfy condition I, the ascription of the function to bring luck to old horseshoes typically does not satisfy condition C, and condition E is not satisfied when one ascribes to light switches the function to detonate accidental spills of gas.

[3] This plan relativity of function ascriptions shows our commitment to an action-theoretic description of use and design. On it, function ascriptions presuppose a rather complicated mental state, more complicated at any rate than states such as 'intending' (conceived as some combination of desiring and believing).

In other respects, the ICE-theory is also more liberal than it may seem. It does not, for instance, require that the account *A* is correct. Even trained engineers are known to use theories that are at best approximately correct; we allow for function ascriptions that are based on such approximations. Furthermore, we do not require that the artefact actually has the capacities that designers and users believe it has. Such a requirement would rule out malfunctioning, which is an important aspect of the phenomenology of artefact use (see also Sect. 2.3).

And this does not exhaust our tolerance. In addition to the central definition, the ICE-theory accounts for another type of functional descriptions of artefacts. This type may arise on the observer perspective, introduced at the end of Sect. 2.1. As said there, an observer constructs a use plan that she/he has neither designed nor learnt about via communication from other agents. There are two possible scenarios in which functional descriptions are based on such a construction. In the first scenario, an observer identifies an artefact's capacity to ϕ as one that contributes to realising a goal, and she/he simultaneously assumes that the artefact has been deliberately designed or used for realising that goal. In this case, the observer may ascribe this capacity as a function to the artefact in accordance with the central definition: she/he assumes that condition E holds, she/he may have the two beliefs required by condition I, and she/he may be able to justify these beliefs as required by condition C.[4] In the second scenario, an observer identifies an artefact's capacity to ψ as one that contributes to realising a goal and simultaneously assumes that it has *not* been deliberately designed or used for realising that goal. This scenario is not amenable to the central definition: the observer only takes the artefact *as if* it is embedded in a use plan – she/he does not truly believe that it is. Thus, condition E remains unfulfilled, although the observer may have the beliefs required by condition I, and these beliefs may be justified as required by condition C. Still, the observer may say that the artefact 'is *functioning as* a ψ-er'. To account for this common usage, we add these watered-down functional descriptions to our theory, labelling them 'ascriptions of functional roles' to distinguish them from 'proper' function ascriptions.

> An agent *a* justifiably describes a component *c* as functioning physicochemically as a ψ-er relative to an item *x* with the physicochemical capacity to Ψ and composed of *c*, *c′*, *c″* … in configuration *k*, and relative to an account *A*, iff:
>
> I. *a* believes that *c* has the capacity to ψ.
> *a* believes that *x* has the capacity to Ψ due to, in part, *c*'s capacity to ψ.
>
> C. *a* can justify these two beliefs on the basis of *A*.

[4] An observer who identifies an artefact's capacity to ϕ as one that contributes to realising a goal, and who assumes that it has been deliberately designed or used for realising that goal, need not always have the two beliefs required by condition I. She/he may, for instance, think that these beliefs are unjustifiable. Take an observer who considers old horseshoes, which are supposed to have the capacity to bring luck. If the observer does not believe her/himself that horseshoes have this capacity, but still ascribes this capacity as a function to the horseshoes, this counts on the ICE-theory as an ungrounded function ascription. The observer can, however, still claim that the people that proposed the 'luck-bringing' use plan for horseshoes, or those that carry out this plan, ascribe the mentioned function to the shoes.

2.3 Assessing the ICE-Theory for Technology

The ICE-theory is normative and based on a partly revisionary analysis of using and designing. As such, it does not simply describe functional descriptions in the technological domain: we describe as designing practices that are regularly not regarded as such, and the theory leaves room for rejecting actual function ascriptions to artefacts as unwarranted. Yet we think that the theory should account for several features of actual functional descriptions. In Vermaas and Houkes (2003) and, in slightly modified form, in Houkes and Vermaas (2010), we lay down four desiderata for a theory of artefact functions. Together, these desiderata significantly constrain the extent to which such a theory can be normative and revisionary.

The four desiderata, all explicitly formulated for the domain of technical artefacts, are:

The proper-accidental desideratum for technology:
: A theory of artefacts should allow that artefacts have a limited number of enduring proper functions as well as more transient accidental functions.

The malfunctioning desideratum for technology:
: A theory of artefacts should introduce a concept of a proper function that allows malfunctioning.

The support desideratum for technology:
: A theory of artefacts should require that there exists a measure of support for ascribing a function to an artefact, even if the artefact is dysfunctional or if it has a function only transiently.

The innovation desideratum for technology:
: A theory of artefacts should be able to ascribe intuitively correct functions to innovative artefacts.

The ICE-theory meets these desiderata in the following way.[5]

Firstly, we can distinguish between proper and improper use plans. Proper use plans are deemed acceptable within a certain community, whereas improper ones are socially disapproved of. Adding our analysis of designing, we can say that, in a community, some designers are acknowledged to be reliable professionals, who typically develop successful use plans; as such, their use plans are socially accepted. Other agents may be involved in designing, but if they are not socially acknowledged as designers, their use plans do not come with an automatic social fiat. This professional division of labour must, of course, be supplemented with other social mechanisms. For instance, a use plan that sprang from the mind of a 'creative user' – who we would call a designer – may gradually be adopted in a community, supplementing or even replacing the original use plan. We do not elaborate on these mechanisms here. For on the basis of the distinction between plans, we can define proper functions as those functions that are ascribed in the ICE-theory relative to proper use plans. The enduring nature of proper function ascriptions stems from the

[5] A more extensive argument that the ICE-theory is meeting these desiderata is given in Houkes and Vermaas (2010).

relative stability of the social acceptance of use plans. Conversely, other function ascriptions to artefacts are transient because they are made relative to use plans that lack social entrenchment.

The malfunctioning desideratum is met by the ICE-theory in the following senses. In cases in which an agent does not know that an artefact does not have the capacity corresponding to its (proper) function or that it cannot exercise that capacity when the corresponding use plan is carried out, the agent can still justifiably ascribe that capacity as function to the artefact: the agent has the beliefs required by the I-condition, can justify these beliefs by an account or by testimonial evidence, as required by the C-condition, and has obtained this testimony by communication from designers, as by condition E. In cases in which an agent does know that an artefact does not exercise the capacity corresponding to its (proper) function when the corresponding use plan is carried out, the agent can assume that this failure is due to that auxiliary conditions of the use plan were not satisfied, but still maintain that the artefact has this capacity[6] and ascribe this capacity as a (proper) function along the lines sketched above. (Yet, in cases in which the agent knows that an artefact does not have a specific capacity, this agent clearly cannot ascribe on the ICE-theory this capacity as a function to the artefact.)

The support desideratum is also met. This is partly a result of the construction of our theory since we require that the capacities that are ascribed as functions to artefacts refer to physicochemical capacities. Nevertheless, we have a more detailed story to tell about the support for function ascriptions. The C-condition ensures that there is evidence that the artefact has these physicochemical capacities. Whatever the source of this evidence, it supports the belief that the artefact has the corresponding capacity as a physicochemical capacity and the belief that this capacity explains, in part, the effectiveness of a use plan. Through the support for these beliefs, the evidence supports the function ascription.

Finally, the ICE-theory straightforwardly meets the innovation desideratum: the historical perspective required to ascribe functions with the ICE-theory may be limited to the design process; it need not extend to earlier generations of artefacts. An artefact can therefore straightaway be ascribed the capacity for which designers selected it.

We have set these four desiderata as conditions for an adequate function theory in the domain of technology. Therefore, we have shown that the ICE-theory is adequate in that domain, just as we have shown elsewhere that other theories, such as existing etiological approaches towards functions, are not (Vermaas and Houkes 2003). Yet assessing the ICE-theory need not stop here because one might want to introduce further desiderata. One reason is provided by the looks of the theory: it takes only a superficial inspection to see that the ICE-theory is an *epistemic*

[6] One illustration that a failure to execute of a use plan need not imply that the relevant artefact lacks the capacity to let this execution be successful is a car with an empty gas tank: that car still has the capacity to drive, although manipulating it by the use plan will lead nowhere. See Houkes and Vermaas (2010) for further argumentation along this line.

function theory, not an 'ontological' one. On the theory, agents arrive at functional descriptions of artefacts on the basis of *justified beliefs* about use plans and about further human beliefs and actions. Agents ascribe physicochemical capacities as functions to the artefacts because these capacities supposedly *explain* why certain goals may be realised by means of the artefact. Thus, functional descriptions of artefacts are ways in which agents highlight, on the basis of evidence, practically relevant capacities. To put the same point more crassly, functions are labels that agents put on supposed physicochemical capacities. The ICE-theory does not entail that these labels refer to real features that exist 'out there', independently of our beliefs. Indeed, the way in which we accounted for the phenomenon of malfunctioning shows that the labels may not refer to real capacities at all. In this sense, the ICE-theory is not an ontological function theory.[7]

A second, related feature of the ICE-theory is that it makes functional descriptions of artefacts dependent on the mental states of both the agents giving these descriptions (through the I-condition) and of the designers that developed the use plans relative to which these descriptions are given (through the E-condition). Thus, if one likes functions to be mind-independent, one has reason to dislike the ICE-theory. To express this sentiment, one might propose a fifth desideratum of the form:

The mind-independence desideratum for technology:

> A theory of artefacts should define functions in such a way that functions do not depend on the mental states of human beings.

We deem it acceptable that functional descriptions of artefacts are derived from the use plans that designers develop for these artefacts; thus, we reject this desideratum. Yet we acknowledge that similar mind-dependence desiderata might be relevant for function theories in other domains – most notably, biology.

3 An ICE-Theory for Biology

We developed the ICE-theory for the technical domain. Still, given the current focus on biology in the philosophy of functions, it is natural to ask how this theory fares with respect to biological functional descriptions. Therefore, in this section, we show how the ICE-theory may be exported to biology. As we said in our introduction, this exporting can be attempted in three different ways: by applying it literally to biological items, by considering biological items *as if* they were technical artefacts and then applying the ICE-theory literally or by exporting the 'ICE-approach' to the biological domain for arriving at an ICE-like biological counterpart of the ICE-theory.

---

[7] The question of whether there exists a theory of technical functions that is compatible with the ICE-theory and that can be taken as an ontological function theory is considered in Vermaas (2009). In Houkes and Vermaas (2004), it is argued that, on such an ontological counterpart of the ICE-theory, technical functions cannot be interpreted as essential properties of artefacts.

3.1 Applying the ICE-Theory Literally to Biology

A literal application of the ICE-theory to biological items quickly runs into problems. The central definition of justifiable function ascriptions presupposes an action-theoretical description of using and designing that is clearly unavailable in biology: according to current neo-Darwinian orthodoxy, there are neither intentional designers of biological items nor use plans for these items that were developed to contribute to goal realisation by other agents. Even creationists may find these assumptions somewhat too steep to accept. For the action-theoretical, use-plan description presupposed by the ICE-theory does not merely imply that artefacts are items that are intentionally shaped by agents. It provides a far richer, utilitarian account: all artefacts are embedded in use plans and are thus means for realising the goals associated with those plans; designers develop those plans and communicate them to other agents, thus creating communication chains between agents who distribute the use plans. Hence, taking biological items as designed or created by a deity is not sufficient to reproduce the use-plan description: one also needs to take the items as designed to be means to ends and to accept that the designer(s) informed the relevant users about these uses.[8]

One may, of course, attempt to put this use-plan description between brackets. Arguably, this bracketing already takes place in the technological domain: when engineers ascribe functions to, for instance, components, they may not do this relative to use plans for those components but relative to capacities of the artefact of which the components are part (Vermaas 2006). But although the use-plan description is bracketed in these function ascriptions, it can still be presupposed. The agent ascribing a function to the component can without contradiction believe that there is a use plan for the component and in that sense still assume that the use-plan description presupposed by the ICE-theory applies. She/he need not know this use plan, but she/he can believe that the agents who are part of the communication chain set up by the designer of the component do know it. One may choose to ignore these points and apply the ICE-definition to biological items anyway. But then the use-plan description is amputated and becomes a ghost limb to the ascription rather than bracketed: the agent ascribing a function to the biological item can now not anymore believe without contradiction that there is a use plan for the item and thus not assume that the use-plan description presupposed by the ICE-theory applies.

Therefore, if the ICE-theory is applied literally to biology, only functional roles may be ascribed to biological items. An agent may identify an item's capacity to ψ as one that contributes to realising a goal and may ascribe this capacity to the item as a functional role. In doing this, the agent does not assume that the item has been

[8] McLaughlin (2001, ch. 7) rejects the analogy between function ascriptions to artefacts and biological items by arguing that artefacts and organisms are associated with different goals, and that designing differs from natural selection. The position that function ascriptions to artefacts presuppose an action-theoretical background that is absent in biology may be taken as a third argument.

deliberately designed or used for realising that goal. Proper function ascriptions to biological items are, in contrast, ruled out, because condition E of the central definition remains unfulfilled.

This puts severe limitations on functional discourse in biology. It becomes, for instance, impossible to distinguish between proper functions and accidental features. Hence, if one introduces a proper-accidental desideratum for biology – as most function theorists since Larry Wright (1973) have effectively done – one has sufficient reason to reject the ICE-theory as inadequate to this domain.

3.2 *Taking Biological Items as As-if Technical Artefacts*

The above results do not rule out that biological functional descriptions can be understood as descriptions that arise by taking, in a Dennettian style, biological items *as if* they were artefacts.[9] In such an as-if description, a fictitious distinction between proper and accidental use plans might be made to avoid the problems discussed above. Let us for the moment assume that such a distinction can be made. Even then, a closer inspection shows that a description of biological items as if they were artefacts carries more detailed and possibly less attractive presuppositions than assumed and/or presented by some authors defending this position. In the literature, the only presupposition that is often considered is that the biological item can be taken as if it were designed by a rational agent in response to problems posed by its biological context. Designing is then taken in its simple, object-oriented sense of intentionally determining the physicochemical structure of the object concerned and not in the action-theoretical sense presupposed by the ICE-theory. For Tim Lewens (2004), who analyses what he calls the 'artefact model' of evolution as 'the approach to the organic world that treats it as though it were designed' (p. 39), the antecedent seems indeed design in this simple sense. Mohan Matthen (1997), who describes the role of the 'product analogy' in biological functional descriptions, takes one step towards our action-theoretical sense of designing, by requiring that users are identified. These users are not agents, but the organisms that benefit from the item to which the function is ascribed: the user of a liver, for instance, is the body that uses it to metabolise fats (p. 31).

Yet the ICE-theory shows that in order for such an as-if description to support function ascriptions, more detailed presuppositions ought to be accepted. Biological items should be embedded in the action-theoretical background discussed above. One has to suppose that the liver has been selected by agents as part of the development

[9] Biological items that are the result of human interferences with biological organisms, ranging from breeding to genetic engineering may be taken as belonging to both the biological and technical domain. Functional descriptions of these hybrid items are ignored here in order to focus on purely biological cases.

of a use plan for the liver. This plan was meant for and communicated to other designers,[10] who are involved in designing whole organisms, such as mice. Mouse designing, in turn, requires a use plan for the mice, which is constructed by designers and communicated to agents who count as users of the mice.

On first sight, these additional presuppositions need not spell trouble for those who think that biological functional descriptions can be understood as as-if artefact-function ascriptions. After all, an as-if description does not need to be realistic in the first place, so no harm seems done by adding a few outlandish brush strokes. But, in the end, these extra presuppositions decrease the plausibility of this account, because – once made explicit – they might make the biologists who describe biological items in functional terms much more reluctant to take these items as if they are artefacts. The users of a biological item can, for instance, no longer be identified as the organisms containing that item, as Matthen wants it: the use plan of a liver is communicated to other designers, not to the body that contains the liver. Lewens' view that the artefact model of evolution 'only becomes practically applicable and psychologically attractive to inquirers' for items that are the result of processes that create 'systems with traits that have the kind of functional complexity reminiscent of designed objects' (Lewens 2004, pp. 119–120) also becomes difficult to maintain. On Lewens' view, the applicability of the artefact model only requires biologists to appreciate the similarity between biological items and designed objects (in Lewens' sense of intentional determination of structure). Then, these items may be described as if they were designed (in Lewens' sense) by a rational agent in response to problems in the item's environment. But this appreciation of similar complexity is insufficient to accept the more detailed presuppositions of the ICE-theory. These require biologists to take biological items as similar to objects that have use plans and that are designed to be used. Prima facie, these latter as-if assumptions are far less plausible and 'psychologically attractive' than the former, more modest ones.

3.3 Constructing a Bio-ICE-Theory

The third route for applying the ICE-theory to biology consists of exporting the 'ICE-approach' to the biological domain, of constructing or selecting a function theory for biology with this approach and of showing that both this theory and the ICE-theory for technology are instances of an overarching theory. Let us call this biological counterpart of the ICE-theory the 'bio-ICE-theory'. Strictly speaking we could select any function theory adequate to the biological domain as the

[10] On the action-theoretical description presupposed by the ICE-theory, the use plan developed for a technical component is primarily developed for and communicated to other designers, who can use the components for designing artefacts according to their own use plans (Vermaas 2006). In the main text, this description of component designing is exported to the liver: as an organ, it is the counterpart of a technical component, rather than a whole artefact.

bio-ICE-theory and define the overarching theory as one that yields the ICE-function when applied to the technical domain and that yields the selected theory when applied to biology. But this manoeuvre would trivialise the unification project. Instead, we are after a bio-ICE-theory that is genuinely similar to the ICE-theory.

One way of transposing a theory from one domain to a similar theory for another domain is by keeping the structure of the theory intact and translating those key concepts that are particular to the original domain into counterparts for the new domain. This procedure is not without ambiguities, nor does it guarantee success. The etiological theories by Millikan (1984, 1993) and by Neander (1991a, b), for instance, are geared to the biological domain but also made applicable to technical artefacts by translating, among others, the biological concept of selection into a technological counterpart. This has not only led to two different translations in the writings of Millikan and Neander – one-shot selection by designers versus long-term selection processes by, e.g. users – but also to function theories that fail to satisfy the desiderata one would like to impose on theories of technical functions (Vermaas and Houkes 2003).

To convert the ICE-theory into a function theory for biological items, one needs to do more than find biological counterparts for the concepts of 'artefact' and 'use plan'. In addition, the use-plan description needs to be transposed, i.e. concepts such as 'designing', 'using' and 'communication' should be given biological counterparts.[11] As a first attempt, we could take use plans as goal-directed patterns *p* in the behaviour of organisms, designing as the process of natural selection of those patterns, communication as the passing on of the results of that selection through genetic information and using as the expression of that genetic information in new organisms. This translation leads to an etiology-style function theory and therefore puts the bio-ICE-theory in good company. This first selection of counterparts, however, is inaccurate. For in the use-plan description, designing and using are not just processes, but also define roles that *agents* play with regard to artefacts. The biological counterparts of using and designing should also define such agent roles, but now with regard to biological items. Furthermore, the two roles defined must be different: designers are at the start of chains of agents who communicate use plans to one another, whereas users lengthen or end these chains.

The following selection of counterparts satisfies these conditions. Let the biological counterpart of designing be the process of *discovering* or *designating* a particular series of behaviours of biological items in an organism as a pattern *p* that is directed towards a specific goal. Let the biological counterpart to the designer be the agent who designates the pattern *p* by identifying the biological items *x* taking part in the pattern and by defining the goal *g* that the pattern is supposed to realise. Call this agent the 'discoverer' *d*. Let the counterpart to designer-user communication be the communication between the discoverer and other biologists who, through

[11] Ulrich Krohs (2009) introduces a generalised concept of design that also makes sense in biology. This generalised concept is, however, not action-theoretical: Krohs' concept of design does not refer to the process of designing but exclusively to the end result of that process. Moreover, Krohs does not introduce equally generalised concepts of using and communication.

this communication, learn about pattern p. Finally, let the counterpart of using be the process of learning about explanatory patterns, for the purpose of better understanding organisms and the behaviours they engage in. This learning process defines – relative to a particular pattern p – the agent role of layperson, which is the counterpart of the user role in technology.

With these counterparts, it is possible to transpose the full action-theoretical description of using and designing of technical artefacts. Instead of the use-plan analysis in technology, we then obtain an action-theoretical description of discovering and learning about goal-directed patterns in which biological items are supposed to play a part. Relative to these patterns, agents may ascribe to a biological item the function to φ if they have justified beliefs that the item has the capacity to ϕ and that the pattern leads to its goal because, in part, the item has the capacity to ϕ. In this bio-ICE-theory, ascribing functions to biological items again means that agents highlight those of the items' capacities that are relevant to understanding the effectiveness of the patterns.

In full detail, we obtain the following central definition of functional descriptions of biological items:

> An agent a justifiably ascribes the physicochemical capacity to ϕ as a function to a biological item x, relative to a behavioural pattern p for x and relative to an account A, iff:
>
> I. a believes that x has the capacity to ϕ.
> a believes that p leads to its goals due to, in part, x's capacity to ϕ.
>
> C. a can justify these two beliefs on the basis of A.
>
> E. a communicated p and testified these beliefs to other agents, or a received p and testimony that the discoverer d who identified p has these beliefs.

3.4 *Assessing the Bio-ICE-Theory*

The bio-ICE-theory constructed above may strike one as bizarre. And it is certainly strikingly dissimilar from the biological function theories currently under discussion in the literature; of all those proposals, it bears only a slight resemblance to John Searle's (1995) theory, in which biological items have their functions relative to goals that agents assign to organisms.

It is, in fact, possible to argue that the bio-ICE-theory is not just dissimilar from existing proposals but also inadequate for biology, if one accepts a desideratum for biological function theories that is largely implicit in the literature. As a counterpart to the ICE-theory, the bio-ICE-theory is *epistemic* and not 'ontological'. Its central definition does not impose conditions for biological items having biological functions. Instead, it states conditions under which agents may single out as functions those supposed physicochemical capacities that they believe to contribute to goal-directed patterns. What is more, these beliefs are not only based on evidence about items, capacities and patterns: they also require evidence about the beliefs of other biologists – the discoverers d who identified the items and designated the patterns. This means that on the bio-ICE-theory, biological functions are mind-dependent: they depend on the beliefs of at least the discoverers *d*. Therefore, the bio-ICE-theory does not meet the biological counterpart of the mind-independence desideratum introduced – and rejected – for technology in Sect. 2.3:

> The mind-independence desideratum for biology:
>
> A theory of biological functions should define functions in such a way that functions do not depend on the mental states of human beings.

For biology, introducing this desideratum makes sense, so it is tempting to dismiss the bio-ICE-theory out of hand. Yet it may pay off to resist this temptation, for the theory actually has some advantageous features.

Firstly, it does not refer necessarily to evolutionary theory. Instead, this theory is one source for biologists to designate behavioural patterns relative to which functions are ascribed; in principle, many other theories and accounts may be used for this. This feature may not be regarded as all that advantageous, given the emphasis that etiological function theories put on evolutionary theory. However, it allows the bio-ICE-theory to make sense of pre-Darwinian and non-Darwinian functional descriptions in biology: before Darwin, biologists may justifiably have employed other sources to designate goal-directed behavioural patterns and ascribe functions; and in domains – biochemistry – in which evolutionary theory plays a more limited role, biologists may also have justifiably singled out goal-directed patterns and ascribed corresponding functions.

Secondly, at least three of the four desiderata that we introduced as touchstones for assessing function theories in technology (see Sect. 2.3) also apply to biology, and it can be argued that the bio-ICE-theory meets these desiderata. To put it very shortly, the biological proper-accidental desideratum is met by distinguishing 'proper' function ascriptions, made relative to patterns that are regarded as characteristic for the organisms concerned, from transient function ascriptions, made relative to idiosyncratic patterns; the biological malfunctioning desideratum is met because the bio-ICE-theory allows for cases in which agents justifiably believe that a biological item has the capacity corresponding to the (proper) function ascribed, even if the item does not have that capacity or cannot execute it; and the biological support desideratum is met because the C-condition ensures that it can be justified that the biological item has the highlighted physicochemical capacities.[12]

Furthermore, even though the bio-ICE-theory is epistemic, it may be possible to formulate a compatible ontological theory about biological items having functions (Vermaas 2009). If such an ontological theory exists, the physicochemical capacities highlighted on the bio-ICE-theory as relevant to behavioural patterns can be taken as real entities that exist 'out there'. Such an associated theory would, however,

[12] It can also be argued that the bio-ICE-theory meets a biological innovation desideratum. But due to the relatively slow development of new biological patterns, the concept of an 'innovative biological item' may not be a relevant one in biology, taking away a basis for introducing this desideratum as a necessary condition for theories of biological functions.

not turn biological functions into mind-independent entities. Compatibility with the bio-ICE-theory would make biological functions dependent on the beliefs of the discoverers d that single out the goal-directed patterns relative to which biological items have their functions. This reference to the beliefs of agents may be taken as a disadvantage, but may also be taken as an asset: if one holds that the concept of biological function is ultimately teleological, the bio-ICE-theory explicitly and straightforwardly identifies the goals related to biological functions. These goals are intentionally assigned to the behavioural patterns that biologists designate in understanding biological organisms.

A final reason for not immediately discarding the bio-ICE-theory is that together with the original ICE-theory, it may be generalised to a function theory for other domains in which we find functional descriptions. We briefly discuss this generalisation in the final section.

4 Conclusion: A Unified ICE-Theory

We started this chapter by introducing the ICE-theory, which we have developed to analyse and assess functional descriptions in technology. We showed that this theory is adequate for technology, in the sense of meeting four desiderata specific to that domain. Then, we argued that the ICE-theory cannot be applied literally to biological items, and we showed how the theory brings to light a series of implausible presuppositions when taking functional discourse in biology as arising from describing biological items as if they are artefacts. Finally, we constructed a bio-ICE-theory, which could be combined with the ICE-theory into a two-domain function theory. On the bio-ICE-theory, functional descriptions of biological items are ways in which agents epistemically highlight those capacities that explain the (successful) realisations of goal-directed behavioural patterns designated by biologists. We pointed out, however, that philosophers typically – albeit implicitly – take biological functions to be features that biological items have independently of agent beliefs and designations. If this mind-independence is introduced as a desideratum for biology, the bio-ICE-theory becomes inadequate.

Biology and technology are not the only domains in which functional descriptions occur; we also find such descriptions in cognitive science, psychology, anthropology, sociology and economics. Therefore, the final step in exporting the original, technological ICE-theory would be to generalise the results obtained so far into a grand unified function theory, which is applicable to all these domains. This unified ICE-theory may have the following form:

> An agent a justifiably ascribes the physicochemical capacity to φ as a function to an item x, relative to a goal-directed pattern p for x and relative to an account A, iff:
>
> I. a believes that x has the capacity to φ.
> a believes that p leads to its goals due to, in part, x's capacity to φ.
> C. a can justify these two beliefs on the basis of A.
> E. a communicated p and testified these beliefs to other agents, or a received p and testimony that the discoverer d who identified p has these beliefs.

It is currently beyond our reach to assess this unified ICE-theory for all the domains just mentioned; we lack, for instance, the experience or intuitions to state the desiderata for function theories in cognitive science. So, at this point, we can merely propose the formulation given above as a conjecture of what a unified function theory may look like. We end with a brief characterisation of the epistemic nature of this unified ICE-theory, by generalising some points made earlier in this contribution.

The unified ICE-theory is epistemic, like the original ICE-theory and the bio-ICE-theory. On it, agents are arriving at functional descriptions of items on the basis of justified beliefs about goal-directed patterns and capacities of those items; they ground those beliefs in evidence about the items and about the beliefs and actions of human agents. In functional descriptions, capacities – e.g. physicochemical, mental or social – are ascribed as functions to the items because these capacities *explain*, relative to the evidence, why the patterns are successful means for realising the associated goals. Thus, through functional descriptions, agents highlight capacities that explain the (successful) realisations of patterns designated by (other) agents. More briefly, in the unified ICE-theory, functions are epistemic highlighters that single out the capacities agents justifiably believe to be relevant.

Finally, suppose it could be shown that this unified ICE-theory is adequate for all non-biological functional domains, just as we showed the original ICE-function theory to be for the technological domain. Then, our conclusion would be that one can arrive at a unified function theory through our third route of exporting the approach of the original ICE-theory to other functional domains only on pain of accepting that biological functions are – despite the implicit assumption of most philosophers working on the topic – mind-dependent.

Acknowledgement Research by Pieter Vermaas and research by Wybo Houkes was supported by the Netherlands Organisation for Scientific Research (NWO).

References

Bratman, M. 1987. *Intentions, plans, and practical reason*. Cambridge, MA: Harvard University Press.

Cummins, R. 1975. Functional analysis. *Journal of Philosophy* 72: 741–765.

Houkes, W., and P.E. Vermaas. 2004. Actions versus functions: A plea for an alternative metaphysics of artifacts. *The Monist* 87: 52–71.

Houkes, W., and P.E. Vermaas. 2010. *Technical functions: On the use and design of artefacts*. Dordrecht: Springer.

Houkes, W., P.E. Vermaas, K. Dorst, and M.J. de Vries. 2002. Design and use as plans: An action-theoretical account. *Design Studies* 23: 303–320.

Krohs, U. 2009. Functions as based on a concept of general design. *Synthese* 166: 69–89.

Lewens, T. 2004. *Organisms and artifacts: Design in nature and elsewhere*. Cambridge, MA: MIT Press.

Matthen, M. 1997. Teleology and the product analogy. *Australasian Journal of Philosophy* 75: 21–37.

McLaughlin, P. 2001. *What functions explain*. Cambridge: Cambridge University Press.

Millikan, R.G. 1984. *Language, thought, and other biological categories: New foundations for realism*. Cambridge, MA: MIT Press.
Millikan, R.G. 1993. *White Queen psychology and other essays for Alice*. Cambridge, MA: MIT Press.
Neander, K. 1991a. Functions as selected effects: The conceptual analyst's defense. *Philosophy of Science* 58: 168–184.
Neander, K. 1991b. The teleological notion of "function". *Australasian Journal of Philosophy* 69: 454–468.
Pollock, J.L. 1995. *Cognitive carpentry: A blueprint for how to build a person*. Cambridge, MA: MIT Press.
Searle, J.R. 1995. *The construction of social reality*. New Haven: Free Press.
Vermaas, P.E. 2006. The physical connection: Engineering function ascriptions to technical artefacts and their components. *Studies in History and Philosophy of Science* 37: 62–75.
Vermaas, P.E. 2009. On unification: Taking technical functions as objective (and biological functions as subjective). In *Functions in biological and artificial worlds: Comparative philosophical perspectives*, Vienna series in theoretical biology, ed. U. Krohs and P. Kroes, 69–87. Cambridge, MA: MIT Press.
Vermaas, P.E., and W. Houkes. 2003. Ascribing functions to technical artefacts: A challenge to etiological accounts of functions. *The British Journal for the Philosophy of Science* 54: 261–289.
Vermaas, P.E., and W. Houkes. 2006. Technical functions: A drawbridge between the intentional and structural nature of technical artefacts. *Studies in History and Philosophy of Science* 37: 5–18.
Wright, L. 1973. Functions. *Philosophical Review* 82: 139–168.

Epilogue

Larry Wright

1.1 Revisiting Teleological Explanations: Reflections Three Decades On

Some time ago, I sketched an etiological account of functions and goals [13] (Wright 1976) that has occasioned many thoughtful criticisms and emendations over the intervening years. In one of the more interesting of these, Tyler Burge has objected to my analysis of functions by suggesting that I seriously mischaracterized my interest as being in that noun ("function") rather than in a certain "pattern of explanation" ([5], Burge 2003, p. 512, fn). In the long retrospect now possible, this does capture something important about my youthful thinking, if perhaps too generously. I originally recognized the contextual resilience of "function" by distinguishing a number of its uses – some explanatory, some not. But I failed altogether to anticipate the gaudy variety subsequent literature would find in its adjectival modification,[1] as well as in other, related grammatical forms, and these developments have increasingly obscured, even to me, the relatively modest insight I had at the time.

My primary motivation had been to articulate the nature and legitimacy of appeal to consequences in what might be called broadly causal explanation, that is, in explanations naturally beginning with because. We sometimes explain the existence or nature of something by appeal to what it results in or conduces to, say it is as it is because of a consequence in this sense. Representing empiricist enthusiasm of the time, Wes Salmon had objected to my giving consequences such a role in an earlier essay, insisting that etiologies could only invoke antecedents. So I set out to explore the contexts in which we do naturally appeal to consequences in response to why

[1] A sample: "proper function," "species function," "token function," "as-if function," "biomedical function," and "fully specified function".

L. Wright(✉)
Department of Philosophy, University of California, Riverside, CA, USA

P. Huneman (ed.), *Functions: selection and mechanisms*, Synthese Library, Studies in Epistemology, Logic, Methodology, and Philosophy of Science 363, DOI 10.1007/978-94-007-5304-4, questions, to see whether and under what constraints such appeal could meet prevailing standards of interest and objectivity.

In starting out in this way, however, I had littered my course with snares, the nature of which I could hardly imagine at the time, for it was natural to think of my topic as teleological explanation. But sorting through the importantly different ways in which explanations naturally appeal to consequences, and relating these to various teleological notions, turned out to involve subtleties scarcely noted in the narrow literature I was then exploring and to implicate issues adequately discussed only in some of the darkest and most difficult texts in our tradition. So although I can hardly do justice to the depth and nuance of those issues within the constraints of this volume, I can nevertheless make some advancements on my original thoughts by saying something about the aspects of them that have survived further reflection and by gesturing at the way in which they must be complicated in order to accommodate the broader concerns that now seem relevant.

The distinction I marked by contrasting functions with goals seems to me still to be of fundamental importance, if a bit more complex than I had realized. And though perhaps no simple terminology would have been ideally felicitous in this role, I noted even in those early ruminations a kind of ironic tension in this choice of vocabulary. The idiom of function works very generally, capturing the explanatoriness of consequences in everything from the action of agents to the structural features of organisms, whereas talk of goal-directedness naturally applies only to behavior and very special behavior at that.[2] By contrast, the very special behavior we recognize as purposive seemed to me then, and seems even more so now, to provide a deeper insight into teleology, that is, to point to the source of its significance.[3]

The source in question is of course agency. And even though I nowhere press the fact that not all goal-directed behavior is that of agents,[4] the dim recognition of their conceptual priority is implicit in the examples I cleave to in illustrating the objective deployment of directedness in our descriptive and explanatory practice: they uniformly involve the purposiveness of agents, from primitive to articulate. This of course is what links the discussion to so much else in philosophy, nearly all of which any particular essay must perforce ignore. What I propose here is to set out some emendations of my original views that now seem to be required on further reflection, but which do not raise the most intractable problems of agency: that is, taking largely for granted the objectivity of our teleological characterizations of and dealings with agents.

[2] Paradigms of the function pattern would be the heart's beating in order to circulate blood or even having kidneys in order to extract waste from the bloodstream; of goal-directedness a paradigm would be the hawk diving in order to catch a rodent.

[3] Over the years, this division of labor has struck me increasingly as rather like that Kant attributes to the contrast between theoretical and practical reason. But at the time, I did not have command of this subtlety and was thus unequipped to press it very far.

[4] The homeostatic behavior of organic or mechanical systems, for instance.

1.1.1 *The Function-Pattern*

That we naturally talk of the function of my shoving a towel under the door (to block a draft) or of your pressing the throttle before starting (to set the choke) obscured from me the fact that what is distinctive of the function-pattern appeal to consequences is precisely what distinguishes it from such cases of ad hoc human purposiveness. The artifacts that provide the function pattern its paradigms involve design and creation in a more robust sense than simply resulting from sentient activity. In their production, volition comes to bear only as part of the general causal backdrop against which items – real or potential – are favored for existence by how they fit into human activity. That is, we typically explain the features of such artifacts as selected for their consequences, quite independently of any individuals actually anticipating them at the beginning of the process or even at all. So the paradigms of this pattern are the stovetop's backsplash and the split-second hand of a stopwatch, the existence and nature of which are explained by appeal to consequences they have or would have in a certain canonical application, and they are explained by them even though the process from which they emerged may have involved various serendipities impinging on the conflicting ideas of many agents none of whom anticipated the precise form of the result except perhaps late in a protracted process.

The complex artifacts of our common experience manifest many instances of this pattern. Motorcycles for years had – and most still do have – separate controls for their front and rear brakes. And the reason for this is that it allows modulating the two brakes independently, contributing materially to the ease and security of their use in commonly encountered circumstances. But the origin of this arrangement almost certainly had nothing to do with this convenience, but rather with the fact that the first motorcycles had only rear brakes, and when greater stopping power became important, the easiest thing to do was to add a separate unit for the front wheel. The arrangement survived, however, because of the dramatic facility subsequently manifested. The market forces and production constraints responsible for the existence and configuration of features on everything from cell phones to SUVs are all rather like this, consisting in the sum of innumerable individual decisions (and lapses) occasionally punctuated by sometimes brilliant recognition of new opportunities in the flux. The process of creating even modestly complex artifacts is in this way exquisitely sensitive to consequences, but rarely in the simple manner of an individual's purposive act. The picture is of a lot of trial and error – including unanticipated feedback – within an only sketchily formulated and flexible project.[5] We are thus provided with such amenities as dedicated refrigerator circuits to minimize the risk of spoiled foods and intermittent wipers for use in light rain.

So far so good. But again, the natural focus on examples like these in which consequence etiologies are of greatest interest hid from me the fact that this analytical

[5] Even the intermediate case in which a single individual creates in a single episode an artifact that we would be inclined to call "designed" (e.g., rigs up a brace for an antenna or frost protection for the garden) is distinguished from the paradigms of simple voluntary action in much the same way.

notion did not engage smoothly with the rest of my canonical vocabulary. Consequences explain in all sorts of ways, only some of which are reasonably characterized teleologically. And making this distinction raises normative issues which I had largely neglected, perhaps because they seemed so clear and uncontroversial. Regardless, correcting the oversight will not be simple: as with so much in philosophy, although the issues resolve transparently in practice, saying anything useful about them involves considerable complexity and sensitivity to nuance.

1.1.2 Virtues

Consequence etiologies are not intrinsically normative: as myriad troubling examples in the literature have clearly demonstrated, the consequence explaining the existence or presence of something need not be recognized as a virtue – as good for anything. Current may be flowing through my body because one consequence of its flowing through my body is to keep me from releasing my grip on a metal post, which would break the circuit. This of course does not in the least undermine the appeal to consequences in explanation, but it does point out a second way in which my vocabulary was obfuscatory. For the distinction between virtues and other explanatory consequences is typically so crucial to understanding what's going on in such cases that the teleological conjunction has come simply to signify the former: the current may flow because of its disabling consequence, but not in order that the consequence obtain. In other words, in the standard context, it is perverse to label such explanations "teleological"[6]: teleological explanations provide consequence etiologies, but not all consequence etiologies are teleological explanations.

Although we will shortly touch (all too briefly) on the way normativity derives from agency,[7] it will be better to develop this point independently of that difficult argument. Fortunately, we may, again uncontroversially I think, take for granted that the normativity underwriting the artifact paradigms of the function pattern is determined, even if sometimes in complicated ways, by the values implicit in the human activities for which an item is designed. In familiar contexts, being able to time several things at once is an objectively good thing, facilitating clean up after cooking is a clear virtue. And their status as virtues in these circumstances plays an essential causal role in the existence and nature of stovetops and stopwatches with these properties. Backsplashes and split-second hands thus have virtue etiologies. So stovetops have backsplashes both because they do and in order to make a cooking mess easy to recover from, similarly for stopwatches.

In this pattern, the explanatorily effective virtue is of course that of the kind of "backsplash" or "split hand," not of any particular instance or application. For the particular use to which my stovetop gets put will have played no role in the

[6] This reservation was often expressed by resisting the label "the function of" for the consequence in question, but, for the reasons mentioned, it is better not to place such weight on this noun.

[7] In the concluding section below, I treat the topic in more deserving detail in the fourth chapter of Wright (forthcoming).

(function-form) etiology of its backsplash; moreover, the virtue need not even be manifested in an instance: a particular stovetop may be part of a display and a watch owned by someone with no interest in intervals. So what is explained teleologically by the function pattern is the property of a kind rather than of an individual. Stovetops (that have them) have backsplashes in order to simplify cleanup: my particular stovetop has one only because it has this virtue in standard application. We arrive at this peculiar-sounding result only because our peculiarly philosophical task is to isolate and characterize a particular explanatory pattern that we unproblematically conflate with others in everyday practice. But, since this may still be difficult to see – I certainly missed it when I first wrote – it is worth pressing the issue from a different direction to as it were locate this result in conceptual space by triangulation.

1.1.3 Erotetics

Because clauses answer why questions simply as a matter of grammar, and so, in order to clauses answer the same questions when the context provides the proper normative constraint. One strand of the erotetic literature encourages us to exploit this fact to articulate significant features of any particular broadly causal explanation by appeal to the contextual presuppositions involved in raising the question to which it responds. Of course, the question most naturally soliciting a function-pattern virtue in the case of stovetops is not a why question but rather something like "What are backsplashes for?" In the context in which it does this, however, it is equivalent to "Why do stovetops (like this) have a backsplash?" (Why is that there?), simply presupposing that it is there because it does some particular good. If this presupposition is false, we reject the question in this form ("Oh, it's not there for any reason"), though we may feel constrained to provide a simple antecedent to account for what now seems a very puzzling phenomenon ("Just an artifact of production, perhaps").

Even when a virtue is etiologically relevant, it is effective only as part of a dense causal matrix any other aspect of which may be selected by the context as the puzzle in need of addressing. This is the issue of the "contrast class."[8] That is, I may already understand the ease of cleanup, but still ask the same question, with a different contrast, eliciting a different explanation:

Why a backsplash as opposed to letting the buyer deal with the problem on his own?
Because customers will pay for the added convenience
Why a backsplash as opposed to splatter-free utensils?
Because splatter-free utensils would be too expensive and difficult to use

These explain the backsplash, compatibly with the virtue etiology, by addressing different puzzles that may be raised by the same question about the same case in various contexts. Without a context in which all but a small number of these issues are

[8] See Chap. 5 of van Fraasen (1980) for the most elaborate discussion of this issue, which was anticipated in Scriven (1966, 1975). Peter Achinstein's remarks on what is "captured" by the words *reason* and *because* in Chap. 3 of Achinstein (1983) are in the service of something like the same point.

already settled, an explanatory request cannot arise, and the notion of an explanation loses its sense: what would then be needed not propositions.[9]

Furthermore, some of these contrasts inevitably articulate the status of an explanatory consequence as a virtue, by providing the conditions in standard application on which it improves. We often gesture at this by making the virtue itself comparative. Modern internal combustion engines have two intake valves per cylinder for better breathing, that is, in order to make it easier for the air-fuel mixture to reach the combustion chamber. Obviously, the comparison is with one valve, not three or six: two would not make breathing easier than three or six. But an improvement on an actual status quo is enough to make it a virtue in this context. That improvement explains the effort required to include two valves in the design and that it is an improvement is adequate to underwrite the teleological conjunction. Again, putting it this way addresses only some puzzles about the configuration, anticipating certain misunderstandings. In another context, "Why two?" may be motivated by the contrast with more not fewer; in which case, other aspects of the matter – details of production, cost, operation, maintenance, and even lack of imagination – would be appealed to in response: all compatibly with the virtue etiology, the availability of which in a standard context is presupposed by these other answers.

What led us here, however, was distinguishing this question and these contexts from another set with which they may be harmlessly conflated in practice but from which they must be distinguished in this discussion. The other question is something like "Why does your stove have a backsplash?" when it may be paraphrased to "Why did you get a stovetop with a backsplash?" This question can have exactly the same answer (to facilitate cleanup after cooking) with very similar virtue-etiological force, but what underwrites its application to the instance rather than the kind is an agent's reason for choosing a particular stovetop, not the selection backdrop responsible for there being such stovetops to choose. Of course, the answers overlap because such choices by individual agents are part of the selection backdrop. But that the function-pattern explanation underwritten by this backdrop is relevant to the kind is apparent again in the possibility of nonstandard application. I may buy such a stovetop even though the tile behind the stove is even easier to clean than its backsplash, thus depriving it of virtue in this case, or I may have chosen it because it's perfect for propping up my spice rack (in order to prop the rack), giving it a wholly different virtue etiology in this application. Neither sort of unorthodoxy in the case affects the kind explanations at all.

1.1.4 Natural Selection

Darwin's innovation, from this vantage, was to notice that function-pattern explanations could also be grounded in a certain organismic history, wholly uninfluenced by explicit agential action or design. The way the routine of organismic reproduction

[9] This is the position of the 3-year-old who responds "Why?" to any answer provided to a previous question: she's not yet ready to play the explanation game.

and inheritance is involved in the "struggle for survival" in the wild provides a causal backdrop structurally homologous to that responsible for the shaping of common artifacts. A certain understanding of this complex arrangement shows the features of organisms to be selected for their consequences in a way strikingly like that in which features of artifacts emerge from the turmoil of design and production. Webbed feet and even details of blood chemistry might be there because of what they facilitate, much like the handle of an eggbeater or the independent brakes on a motorcycle: the development of a species being the flexible project benefitting from trial and error and feedback.

The objective virtue of a characteristic would naturally be the advantage it provides in the struggle, and the structural parallel extends to this virtue – and hence the virtue etiology – attaching directly to the species (kind) and only indirectly or figuratively if at all to the individual (application). For the particular use to which Mrs. Mallard puts her webbed feet, no matter how important to her, played no role in her showing up with them, and of course many adaptations of reproduction and rearing, for instance, have the same sort of virtue etiology but actually disadvantage individuals in their own struggle, benefitting only propagation of the kind. So just as with split hands and backsplashes, we can say that ducks have webbed feet in order to facilitate their swimming, whereas Sonya has webbed feet only because they facilitate swimming. When we are inclined, as we are, to say that Sonya has webbed feet in order to facilitate her swimming, we may be thought of as engaged in synecdoche.

Here too, the context may require other, mostly nonteleological answers to the question "Why P?" (Why do ducks have webbed feet?), taking the function pattern for granted. Why webbing – as opposed, for instance, to some other way of exploiting water (such as paddle wheels or water jets as used by the waterfowl on an interlocutor's home planet) – would require appeal to details of developmental and structural possibilities in terrestrial animals or perhaps accidents of their history?[10] Or the question may express a puzzle about the waterfowl niche itself: why it was exploited at all, or in this way, as opposed to developing winged critters with feet better adapted to taking refuge in trees.

The artifact parallel extends to the easy conflation of distinct teleological requests that have generated selfish-gene puns. Many adaptations in animals involve purposive dispositions. And while those dispositions – to stalk, attack, seek, flee, hide, mate, nurture, husband, and the like – have themselves a function-pattern teleological explanation, instances of their manifestation do not, the inevitable synecdoche directing us back to the kind. Just as with her having webbed feet, when Sonya uses them to paddle around an obstacle, the backdrop of natural selection licenses only a because not an in order to. In the right context we can exploit natural selection to say she paddled – or more generally took flight, poked around, etc. – because it conduced to a consequence, the context making it clear that the conducing in question concerns the disposition of which "it" is an instance,

[10] Ron Amundson (1994) describes usefully detailed biological examples in which context sorts through such causal factors rather as it does in the valve-train example of artifact development.

which can be explained only as a virtue to the species, not Sonya,[11] and which will not involve the particular obstacles or predators she is now engaged with. That is to say, the selection etiology explains the instance only indirectly, as a simple consequence of the disposition.

On the other hand, we are right to think Sonya paddled in order to reach shore and the rabbit ducked into its warren for the sake of refuge. Such instances do have a teleological explanation – a virtue etiology – but as actions of a primitive agent, not as the result of the selection backdrop. This much is implicit in attributing the activity to an individual, saying it was what paddled or took refuge. And the independence of the two explanatory regimes is clear from our ability to deal purposively with our pets and other common animals without knowing anything much about the species to which these individuals belong or about the nature of the selection backdrop responsible for their talents. So the parallel with stoves and stopwatches is exact: we must distinguish in analysis overlapping patterns we harmlessly conflate in practice. When we say a critter acted (paddled, bolted) for the sake of an end (to reach shore, take refuge), we may be speaking literally, giving the particular virtue-etiology responsible for the action in question, or we may be speaking figuratively in tracing the very existence of such a marvelously complex disposition to its phylogenetic roots.

Nevertheless, the conceptual convolutedness of agency, and the legendary struggle our tradition has consequently had in articulating its nature and significance, has made it more difficult to pry these two patterns apart here than it is in the earlier example of artifacts. Let this narrowly motivate a concluding look at the goal pattern.

1.1.5 Purposiveness and Agency

My original appeal to agency in exercising the goal-directedness pattern (Wright 1976, Chapter 2) was in large part designed to display the objectivity of behavioral consequence etiologies, and for those wholly unmoved by those efforts, the topic will require far more than a short essay to address. To the extent that they represent a promising start, however, we may make some progress on the matter by explicitly flagging their repressed normative component, as we have in the function pattern, to show the way in which its grounding too is naturalistic.

The objectivity I claimed for our appreciation of primitive agency is found in the intersubjective recognition of things like chasing, fleeing, dodging, struggling, noticing, ignoring, and resting, and distinguishing them from exploding, welling, wafting, falling, boiling, and being simply inert or fixed or dead. What distinguishes the former was characterized by twentieth-century empiricists[12] as its "plasticity" and "persistence" and is evident, even to the novice, in magnified pond water. The point of my earlier

[11] Again, dispositions to engage in risky mating or rearing behavior, which are not clearly good for the individual, will have the same function-pattern virtue etiology.

[12] See Braithwaite (1953) p. 330, ff. and Nagel (1961), p. 410, ff.

ruminations was to show that what we recognized here had the form of an explanation and one that made the behavior dependent on its consequences in a fairly straightforward way. Simply missing an obstacle does not count as dodging it unless the path taken was somehow affected by the fact that it would avoid the object in question. Running behind something is not following it unless its details are a function of its prospects.

My central concern in the second chapter of Wright (1976) was to spell out the objectivity of this perception's appeal to consequences by fleshing out a suggestion of Charles Taylor's, using the form of Mill's Methods. That form articulates the peculiar complexity we recognize paradigmatically as agential behavior in a way that reveals it to have the structure of an experiment: one designed to discover the consequences responsible for the complexity. What stands out in the flux of data is the systematic variation of a pattern's detail with the conditions affecting the significance of that detail. We thus distinguish its locus, the agent, from the neighboring detritus as a source of activity, that is, as a creature doing something in the robust sense of that term.

Although I did not originally understand it in these terms, part of what makes agency fundamental is that, unlike that in the function pattern, the normativity underwriting the teleological conjunction (in order to/for the sake of) is not extrinsic to the explanation, but is implicit in its structure. The resulting fusion of causal with normative issues makes this very tricky to see, however, and is, I think, largely responsible for the expository difficulty associated with this topic over the millennia.[13] And although nothing short of a curriculum can adequately address it, we have room here to at least hint at the way the perspective of this chapter advances on or at least consolidates the gains of that conversation.

In distinguishing between fleeing and wafting and jumping and falling, we are distinguishing behavior that can be attributed to an agent, as something it robustly did, "on purpose," from other things that simply happen to it that are none of its doing. That a pattern in something's behavior may have the explanatory complexity required to identify it as the source of activity – of behavior we may call its action – is responsible both for the normativity naturally generated by these explanations and the grammatical illusions to which they are so deeply subject.[14] We may gain useful insight into this convoluted network of notions, while avoiding the worst of the grammatical hazards, by extending my earlier discussion of primitive agents and noting the empirical credentials of the normativity that emerges at this level.

An agent's action, what it does, on purpose, is behavior that takes place for the sake of something and hence has a virtue etiology. The good making the etiology

[13] I have in mind primarily the conversation extending back at least to Aristotle's *De Anima* and reaching its twentieth-century apotheosis in Heidegger's *Being and Time*, though the tough going extends as well too much of the recent analytic literature on action and agency.

[14] These appear perhaps most explicitly in discussion of the first-person pronouns in the *Paralogisms* of Kant (1781/1997), which Heidegger picks up and embellishes at the beginning of Division I of Heidegger (1927/1962), leading to his giving agency a distinctly adverbial cast, which in turn doubtless influenced Ryle's similar treatment of mental terms in general (Ryle 1949, especially Chapters IV and X). I address these in yet another way in Wright (forthcoming).

virtuous is of course always a function of what may be called, even in primitive agents, their interests or values: something is a good thing to do only if it in fact conduces to the interests of the agent doing the behaving. But these interests are an empirical matter: in explaining a rabbit's activity, we may begin with a certain view of its utilities but be forced to revise it in light of the data we have to account for. We can simply discover, wholly against expectations, that it is not afraid of us, likes blue things, and is fiercely protective of a carrot it managed to find. Details of the activity (patterns of persistence, plasticity with respect to obstacles and interventions) might simply not make sense as flight (done in order to escape), but seem pretty obviously occurring for the sake of supervising its carrot or retrieving a distinctively blue rock from a pile of gray ones.

The crucial characterizational problem is highlighted by the fact, implicit in the empiricists' criteria, that the most strikingly objective cases of purposive behavior, like the rabbit's digging at a blue rock, are not immediately successful: we easily identify behavior as done by an agent independently of its achieving the end toward which it is palpably directed. Prototypical are attack that merely frightens, unproductive foraging, and doomed flight. But for these even to be unsuccessful – failures, attributable to the agent – they must be its action and hence have a virtue-etiology. This is why the verb to try plays such a deep conceptual role in agency. For the notion of attempt allows us to register what is successful in a failure: to say what it was about a foiled gambit that made it a good thing to do – to try – anyway. And of course quite generally, understanding an agent's action is understanding what it is trying to do.

The refractoriness distinctive of agency derives in large measure from the need to introduce yet another parameter to deal with this complication. For whether a particular maneuver is a good thing for an agent to try in a given circumstance depends not just on its interests and values but also on what we might call its "competence": its general level of ability to pursue its ends. In common conditions, a rabbit's bolting when attacked by a dog is not just a good thing to do, but its very best option, even if the dog is faster and quickly overtakes him. But this evaluation would be objectively wrong if the rabbit could have easily climbed a nearby tree or simply pulled a gun and shot the dog. The inextricability of causal from normative issues in articulating our recognition of agency lies in the systematic interdependence of these two components of what we recognize: we require some guidance from ends in order to assess competence and vice versa.

In the event (examples like those above), we do of course have plenty of guidance from each: agential patterns show up best in a rich flux of data and only to observers intimately familiar with agents of the kind in question. The less of either (data, familiarity) we have, the more tentative and experimental is our identification.[15]

[15] The requirement of finite antecedent plausibilities does not distinguish this from any other explanatory diagnosis. As the Bayes formula makes explicit, the suspicions we begin with determine the relevance of evidence. And as with any such suspicions, those about interests and competence are empirically corrigible: we may discover a beast's bad eyesight or peculiar interest in blue things, just as we may discover that it was a not an empty tank, but a plugged fuel filter that stalled the car just short of the filling station.

In every case, however, our recognition of an agent involves diagnosing ends and ability together as a hermeneutic package. The pattern we identify with agency is thus, to slightly distort Robert Brandom's felicitous phrase, a "locus of … responsibility" (Brandom 2002, p. 217). What we recognize and adjust our understanding of in these cases is (a) something responding as conditions affect its interest in a way made appropriate by its array of skills. This is just to say it does the right (appropriate) thing because it is the right thing to do. And since the twin parameters of purpose and proficiency are what we constantly test and modify in our temporally extended observations and interventions, this naturalizes the normativity of agential virtue etiologies. And it is this normativity that ultimately underwrites the teleological conjunction in the function-pattern paradigms.

References

Achinstein, P. 1983. *The nature of explanation*. New-York: Oxford University Press.

Amundson, R. 1994. Two concepts of constraint: Adaptationism and the challenge from developmental biology. *Philosophy of Science* 61(1994):556–578.

Braithwaite, R. 1953. *Scientific explanation*. Cambridge: Cambridge University Press.

Brandom, R. 2002. *Tales of the mighty dead: Historical essays in the metaphysics of intentionality*. New-Haven: Harvard University Press.

Burge, T. 2003. Perceptual entitlement. *Philosophy and Phenomenological Research* LXVI: 503–548

Heidegger, M. 1927/1962. *Being and time* (trans: Macquarrie and Robinson). London: Harper and Row.

Kant, I. 1781/1997. *Critique of pure reason* (trans: Guyer and Wood). Cambridge: Cambridge University Press.

Nagel, E. 1961. *The structure of science*. New-York: Harcourt, Brace and World.

Ryle, G. 1949. *The concept of mind*. Chicago: University of Chicago Press.

Scriven, M. 1966. Causes, connections, and conditions in history. In *Philosophical analysis and history*, ed. W.H. Dray, 238–264. London: Harper and Row.

Scriven, M. 1975. Causation as explanation. *Nous* 9: 3–16.

Van Fraassen, B. 1980. *The scientific image*. Oxford: Clarendon Press.

Wright, L. 1976. *Teleological explanations*. Berkeley: University of California Press.

Wright, L. (forthcoming). The Concept of a Reason.